THE ESSENTIAL
CRYSTAL
HANDBOOK

THE ESSENTIAL
CRYSTAL
HANDBOOK

ALL THE CRYSTALS YOU WILL EVER NEED

FOR HEALTH, HEALING & HAPPINESS

SIMON AND SUE LILLY

DUNCAN BAIRD PUBLISHERS

LONDON

The Essential Crystal Handbook
Simon and Sue Lilly

First published in the United Kingdom
and Ireland in 2006 by
Duncan Baird Publishers Ltd
Sixth Floor, Castle House
75–76 Wells Street
London W1T 3QH

Conceived, created and designed by
Duncan Baird Publishers

Managing Editor: Kirsten Chapman
Editors: James Hodgson and
  Kelly Thompson
Managing Designer: Manisha Patel
Designer: Gail Jones
Commissioned photography:
  Matthew Ward

A CIP record for this book is available
from the British Library

ISBN: 978-1-84483-233-0

10 9 8 7 6

Typeset in GillSans
Colour reproduction by Scanhouse,
Malaysia
Printed and bound in China by Imago

Contents

Introduction

There is nothing more down-to-earth than the rock beneath our feet. So why are gems and crystals regarded as special? Why have our ancestors, as far as history and archaeology can tell, always considered them to hold magical power?

Even today in the 21st century the fascination with crystals continues. Indeed, the interest has blossomed over the past 30 years, matching the spread of mind, body and spirit stores. Even those who are not persuaded by the new spirituality may find themselves searching through baskets of tumbled stones, in the hope of finding a lucky birthstone. At the other end of the scale are bright-eyed seekers who think nothing of spending a week's food bill on a rare configuration of the latest crystal in order to harness its healing or spiritual benefits.

Conventional medicine has relied for its development on the physical qualities of a great range of minerals, with tons of crystals being mined, refined, ground up, cut and transformed into other substances. The scientific community, however, looks with haughty derision, if at all, at the claims that crystals can affect wellness simply by being placed for a few minutes on or around the body.

We are brought up to believe in a rational, scientific world-view, in which everything must make sense, seem plausible, be explainable. To be "educated" means to be pragmatic, rational and sensible.

The new spiritualities of the West, along with alternative therapies, often attempt to justify themselves by adopting scientific-sounding explanations in order to fall in line with the intellectual mainstream. They largely fail in this for one simple reason – at their basis is the unique, personal experience of the individual. "Personal experience" is not a scientifically measurable phenomenon. It is, however, one of the main factors governing each human being's actions, choices and behaviour. Experience is the food of life and we all look for foods that we find tasty and fulfilling, avoiding those that leave us hungry in some way.

Crystals and gemstones have always served this hunger well. Their very attractiveness and beauty is often felt as a mouth-watering taste – our reactions identical to spying a delicious-looking cake or pastry.

This book cannot tell you why crystals seem to be effective healing tools, because no one really knows yet (in a scientific way). What we will show you are some of the main crystal-healing ingredients and recipes for you to try out for yourselves.

As trainers of professional crystal therapists for well over a decade, we have seen time and again how the most straightforward of techniques, using only a handful of small crystals, sometimes not even coming into physical contact with the body, can result in profound improvements in health and well-being.

A trained crystal therapist often has many methods of assessing and correcting energy imbalances in his or her client. These skills focus and fine-tune the healing effect to achieve the maximum benefit. The suggestions offered in this book are necessarily of a more general nature, but nonetheless will provide you with real opportunities to release the underlying stresses that predispose us to health problems.

Part One:
The Power of Crystals

Crystals are incredibly versatile. They can be used in a wide range of ways to improve your life physically, mentally, emotionally and spiritually. This part of the book presents a selection of straightforward healing techniques for you to explore. The measure of how effective crystals are in your life should not be gauged by the experience of a single healing session. On some days you may be more sensitive to the processes than on others. It's always best to have short, regular sessions than to try to "fix" everything all at once. Over the weeks you'll soon begin to notice that your well-being improves, that you cope better with difficult situations, and that you feel more settled and balanced in yourself.

Healing traditions

Crystals have always been considered special, extraordinary, even supernatural. With their fascinating forms and enchanting colours, nothing else on Earth is like them. Ancient peoples believed that crystals had fallen from the heavens as gifts from their ancestors. This notion is not so far from the truth. All the particles that make up matter once originated within the hearts of old stars. When these stars exploded, their particles were scattered throughout the universe and re-formed as new stars and planets.

People who carried, wore or used crystals believed that they would be imbued with some of the power belonging to these spiritual realms. Religious and temporal leaders would seek out the finest gems. The general populace had no access to such rare and expensive items but would mimic their appearance with less precious materials, hoping for some benefit. In ancient Egypt, for example, the pharaoh and his priesthood prized turquoise as a symbol of fertility, life and joy, while the people imitated this stone with amulets of blue glass.

In this way the two main areas of crystal use were defined: as a means to display earthly status and as a focus for magical and healing energies. These two strands persist to this day.

Found on the mummy of the pharaoh Tutankhamun (reigned c.1336–c.1327BC), this gold pectoral ornament contains lapis lazuli, calcite and turquoise.

From ancient to modern

The forms of crystal healing that are in use today arose in the alternative cultures of the 1970s and 1980s, but like many complementary systems, they have developed out of old cultural traditions. The idea that every stone has special magical properties can be traced back through European literature of the early Middle Ages to their sources in Greek, Roman and Arabic philosophy, which in turn owe much to the Ayurvedic traditions of India.

Modern crystal healers initially focused very much on the properties of rock crystal (clear quartz). Traditionally crystals of quartz have been highly valued by healers around the world for their ability to connect with spirit worlds, to reveal and diagnose illness and to allow a shaman or healer to remove the causes of disease. The use of crystals, crystal mirrors and crystal balls for divination is simply a continuation of this long tradition.

These mystical ways of using crystals are usually dismissed by modern science as mere superstition, and yet science recognizes the unique – and often very strange – properties of minerals and crystals, making use of their characteristics in everything from wristwatches (quartz), automobile spark-plugs (kyanite), medical lasers (ruby) and space shuttle windows (sapphire).

The pragmatic and metaphysical views of the crystal world may seem to be mutually exclusive. However, both approaches are valuable. One provides useful tools and technologies, while the other enables us to see beyond the here and now into the realm of imagination and potential. If we only pay attention to one of these sides of our natures, we miss out on a wealth of experience – both are necessary to keep us whole and healthy. Crystal healing can restore fundamental balance to our existence.

Crystals and colour

Qualities like chemical make-up, structure, shape, hardness and lustre focus the energies of crystals in certain ways, but colour is the key to how a crystal heals.

We appreciate colour as an adornment, forgetting that it represents a real force – the level of energy in light. This energy stimulates the brain, altering the way we think and feel, as well as our physical well-being. Our reaction to colour is instinctive and cannot be controlled. But if we learn the effects of each colour, we will begin to understand why we are drawn to some stones more than others, which will help us to choose the right crystal for every situation.

Many crystals appear in more than one colour. This is because some samples may contain a few extra atoms of an uncharacteristic element that slightly distorts the internal lattice and changes the colour we perceive.

Gem experts discover the true colour of a crystal by a streak test. This involves scraping the crystal across an unglazed ceramic tile to leave a fine dust of particles, revealing the crystal's underlying colour. The streaks of smoky quartz, violet amethyst, yellow citrine, rose quartz and green aventurine, for example, are all grey-white – exactly the same as for clear quartz.

Basic colour qualities

Black	grounding	Blue	communicating
Brown	enabling	Indigo (deep blue)	cooling
Red	energizing	Violet	integrating
Orange	releasing	Pink	accepting
Yellow	organizing	White	cleansing
Green	balancing	Clear	clarifying

The aura

All living things have an aura – an energy field that surrounds the body. The aura is made up of several layers, which we cannot detect with our physical senses, but only using inner awareness.

Each layer of the aura contains certain aspects of our make-up. A problem in one layer will often affect other layers until the imbalance spreads through the aura and the body, and manifests as distress or disease. Crystal healing can quickly clear away these subtle blemishes in the same way that regular window cleaning allows maximum light to enter a room.

The layers of the aura are often represented like the concentric skins of an onion. In fact, the most subtle outer layers interpenetrate all the others, and they all weave together within the physical body.

Levels of the aura

Etheric body Closest to the physical, this acts as a blueprint for our organic body, and shows imbalances just before they manifest on a physical level.

Emotional body This level is seen by psychics as swirls of colour that show our ever-changing emotions; it is easily cleared and balanced by crystals.

Mental body A structure that holds our ideas, beliefs and experiences.

Stress here can lead to rigidity and prevent us from achieving our goals.

Astral body A boundary between everyday awareness and spiritual nature. Disorientation, fear and paranoia are symptomatic of imbalances here.

Spiritual bodies The subtle outer levels are more difficult to define. They feed us with universal energy, which helps us to reach ultimate fulfilment.

13

The subtle body and chakras

Originating in ancient India, and adopted by many alternative healing systems (including crystal healing), is the concept of the subtle body, which shares many of the principles of the aura (see page 13). Linked to, but transcending, the physical body, the subtle body is a scientifically undetectable network of channels (*nadis*), through which flows life-energy (*prana*). Our physical, mental and spiritual well-being depends on *prana* flowing smoothly to all parts of the subtle body. If flow becomes sluggish, owing to a congested channel, or, on the other hand, too vigorous, then the likely result is physical illness or mental or spiritual unease. Identifying and correcting inappropriate energy patterns helps the body to regain balance, release stress and repair damage.

We can optimize the flow of *prana* by influencing the main receptors and distributors of life-energy in the subtle body, which are known as chakras (from the Sanskrit, "wheel"). The seven principal chakras are located in a line running down the centre of the body from just above the top of the head to the base of the spine. Each of these chakras governs certain physical, mental and spiritual aspects of our being, and when functioning well, each chakra is believed to vibrate at the frequency of one of the colours of the rainbow spectrum (see diagram, opposite). A malfunction in one chakra affects all the others, so it is important that the whole system is in balance.

Balancing chakras is at the heart of crystal healing. We can use crystals to "feed" each chakra with its main energetic colour. At the simplest level, placing one stone of the appropriate colour on each chakra for a minute will help to balance the whole system. This is by no means the only way to work with crystals and chakras, but it will be helpful in most cases.

Crown chakra (violet)
Just above top of head
Integration, coherence,
sense of belonging and
spiritual ease

Brow chakra (indigo)
Centre of forehead
Sight, comprehension,
perception, sense
of perspective, vision
and intuition

Throat chakra (light blue)
Base of throat
Communication,
expression, energy
exchange, peace and
understanding

Heart chakra (green)
Centre of chest
Relationships, balance,
space, growth, love,
affinity and calmness

Solar plexus chakra (yellow)
Between navel and diaphragm
Personal power, control,
understanding, digestion,
sustenance, integrity and
confidence

Sacral chakra (orange)
Below navel
Creativity, exploration,
sensation, pleasure,
sensuality, flow, repair
and enjoyment

Root chakra (red)
Energy, motivation,
groundedness, practicality,
survival needs and
connection to the Earth

*(Note: the root chakra is at the
base of the spine, but it is often
balanced by placing a crystal
between the knees or thighs)*

15

Ayurveda

Having existed in India for at least 2,000 years, Ayurveda (Sanskrit, "science of life") is one of the world's oldest systems of medicine. A holistic tradition that integrates massage, exercise, diet and meditation, Ayurveda accords a special role to gemstones. According to ancient texts, the light of the Creator is transmitted to Earth by the planets; gemstones collect and then radiate this energy. If a person's well-being is affected by inharmonious influences, an ayurvedic healer can identify which planet is responsible and prescribe the appropriate gem to wear in order to correct the disruption.

The list below shows, for each ayurvedic planet, the aspects that it represents and the principal gemstone associated with it.

Planet and gemstone correspondences

Sun (male aspects, vitality, success) – **ruby**

Moon (feminine aspects, peace of mind, comfort, well-being) – **pearl**

Mercury (education, intelligence, speech, humour) – **emerald**

Venus (love, beauty, art) – **diamond**

Mars (courage, confidence, physical strength) – **coral**

Jupiter (philosophy, spiritual growth, wealth, learning) – **yellow sapphire**

Saturn (fame, longevity, sincerity, justice) – **blue sapphire**

Rahu, the ascending node of the moon (restlessness, worldly success, force, selfishness) – **zircon**

Ketu, the descending node of the moon (passivity, spirituality, non-attachment, selflessness) – **cat's eye chrysoberyl**

Birthstones

The link between heavenly bodies and gems is continued in the concept of birthstones. Birthstone lists vary because some are arranged according to the zodiac and some according to calendar months. The list presented here is an abbreviated amalgamation of the most authentic traditions. However, it is always more beneficial to use a stone to which you are spontaneously attracted, rather than one you feel you should have simply because it is linked to your zodiac sun sign, which is only one aspect of your full birthchart.

Aries (Mar. 21–Apr. 19) – carnelian, bloodstone, jasper, diamond

Taurus (Apr. 20–May 20) – tourmaline, tiger's eye, topaz, emerald

Gemini (May 21–Jun. 20) – aquamarine, citrine, chrysocolla, tiger's eye, pearl

Cancer (Jun. 21–Jul. 22) – moonstone, pearl, ruby, emerald

Leo (Jul. 23–Aug. 22) – ruby, heliodor, sunstone, clear quartz

Virgo (Aug. 23–Sep. 22) – sapphire, peridot, carnelian, citrine

Libra (Sep. 23–Oct. 22) – lapis lazuli, opal, aventurine, peridot

Scorpio (Oct. 23–Nov. 21) – obsidian, Herkimer diamond, topaz, aquamarine

Sagittarius (Nov. 22–Dec. 21) – blue quartz, amethyst, malachite, turquoise

Capricorn (Dec. 22–Jan. 19) – jet, black onyx, clear quartz, black tourmaline

Aquarius (Jan. 20–Feb. 18) – amethyst, sapphire, chalcedony, garnet

Pisces (Feb. 19–Mar. 20) – turquoise, pearl, rose quartz, aquamarine

17

The geology of gems

Minerals are inorganic materials that form naturally in the Earth. Minerals of the same type share the same chemical formula (except for impurities) and the same ordered crystal structure. They are the most stable form of matter in the universe.

Crystal formation

The mechanism that creates crystals is movement in the Earth's crust. As layers of crust float on the liquid rock mantle beneath, collisions and pressure create fractures that allow superheated liquids and gases, filled with different elements, to move rapidly toward the surface. As these liquids and gases cool, they often crystallize. The temperature, pressure, surrounding rocks and mix of elements all determine which minerals crystallize at a particular place. Crystals and rocks produced in this way are known as igneous.

If crystals and rocks are exposed to wind, frost and water, they begin to erode. Small particles are washed into the sea where they form silt deposits. After many millions of years of extreme pressure, these deposits become compressed, forming sedimentary crystals and rock. Minerals of this type tend to be softer than igneous crystals as they form at lower temperatures and pressures.

Volcanic eruptions bring the fiery processes of igneous crystal formation dramatically to the surface (see opposite). The healing stone obsidian is created from rapidly cooling volcanic lava.

The third mode of crystal formation is known as metamorphic. This is where igneous or sedimentary minerals have been re-subjected to heat or pressure in the Earth's crust. Although the particles do not melt, they may re-combine to form different crystals.

As the Earth's crust is continually moving, these crystallization processes are always in progress. We only see the results when veins or beds of crystals are exposed by erosion or are lifted near enough to the surface to be mined.

Crystal systems

Crystals can remain unchanged for millions of years because their constituent atoms are arranged in a stable, regular pattern, called a lattice. This lattice repeats throughout the crystal and ensures that every crystal of the same mineral will show the same geometrical characteristics, with the same number of flat faces always meeting at the same angles. This precise repetition is the same no matter what size the crystal or whether it is a faceted gem or a dull, rough piece. A crystal's external form is simply a reflection of its internal atomic structure. These shapes have a significant influence over the magical and healing properties of a crystal.

The lattices fall into seven major types (see list, opposite). Different minerals that share the same core pattern will have similar, basic shapes. Unfortunately, these simple forms are seldom obvious from the external appearance of crystals owing to the variable conditions in which they form. However, careful scrutiny will allow you to recognize certain identifying traits (for example, the hexagonal cross-section of quartz or emerald).

Crystal lattice patterns

Cubic

A cube-shaped lattice. Most crystals of this type consist of a number of interpenetrating cubes, rather then single cubes.
Examples: garnet, pyrite, sodalite, diamond, halite, fluorite and copper
Qualities: releasing tension and encouraging creativity

Tetragonal

This lattice is shaped like two pyramids joined base to base, with elongated upper and lower points.
Examples: apophyllite pyramid, zircon, rutile and chalcopyrite
Qualities: balancing and harmonizing

Orthorhombic

Shaped like a small squashed matchbox, with all sides unequal in length.
Examples: danburite blade, peridot, topaz and celestite
Qualities: linking, aiding the flow of information

Trigonal

This lattice is shaped like a diamond or a barrel-shaped lozenge.
Examples: sapphire, alpha quartz, tourmaline and ruby
Qualities: energizing and anchoring

Monoclinic

Shaped like an elongated squashed matchbox.
Examples: selenite wand and kunzite
Qualities: affects movement and perception

Hexagonal

Shaped like a hexagonal prism.
Examples: aquamarine, beta quartz, emerald and apatite
Qualities: organizing and supporting

Triclinic

This lattice is the most variable, having no set angles or lengths.
Examples: kyanite blade, labradorite and ulexite
Qualities: opening and protective

Your crystal collection

Every year hundreds of minerals are discovered. Most are of little interest to the crystal healer as they are often too delicate, rare, small or toxic to be of practical use. In fact, the proportion of crystals used by healers through the ages is tiny compared to the wealth of minerals within the Earth's crust.

Availability, durability and safety, as well as effectiveness and versatility, should be considered when collecting stones and crystals for healing. Pay attention to your intuitive preferences, but make sure that you keep a broad-based selection.

Crystals are marketed like any other consumer goods, and common minerals are renamed to capture the imagination. Not all commercial names relate to the true mineral name – for example, "turquentine" is actually dyed howlite, and bears no relation to turquoise. When buying stones, always try to check their mineralogical names, so you know exactly what you are getting. It is easy to mis-identify and mis-label crystals – and store owners can make mistakes – so take time to learn crystals' characteristic shapes and colours.

What to look for

Whatever the size or outward appearance of a stone, it is the internal structure, its lattice (see pages 20–21), that makes the crystal a unique healing tool. It is therefore not important whether a stone is large and flawless or small and battered. The visible shape and geometry of a stone has nowhere near as much effect as its invisible atomic lattice.

Common forms of crystal

- **Mineralogical samples** – crystals or groups of minerals still on their matrix (rock base) or in the form in which they were found.

- **Crystal clusters** – a number of crystals sharing a common base. Quartz clusters, particularly amethyst beds, are popular. They can keep the energy in a room positive and cleanse other stones.

- **Single crystals** – recognizable by their geometric shapes, single crystals have flat, angular sides and faceted ends (called "points" or "terminations"). The base is usually rough where the crystal has been separated from its matrix.

- **Tumbled stones** – lesser grade or damaged crystals can be polished with gravels until they resemble rounded, water-worn pebbles. This enhances their colour. They are relatively inexpensive and easy to place on the body, so form the core of any crystal healer's kit.

- **Worked crystals** – these are shaped and polished from large blocks into massage wands, spheres, free-standing polished crystals and slices, which may be decorative or used in healing. Faceted gems and cabochons (dome-cut stones) as used in jewelry can also be of use.

23

Building a healing kit

For healing try to collect a wide variety of mineral types in a broad range of colours. Tumbled stones, flat stones and small, individual crystals are easier to rest on the body. A basic collection should include:

- around 12 small to medium (2–3cm/ 1in long) clear quartz crystals
- three to four crystals each of amethyst, smoky quartz and citrine
- one or two massage wands, small
- spheres or other polished stones
- one or more crystal pendulums – clear quartz or amethyst work well
- a few larger crystals for meditation – they are easier to hold

Caring for your crystals

Although they may seem robust, crystals need careful handling.

- Soft stones can be abraded by harder stones, so keep them apart.
- Look out in hardware and fishing stores for light, compartmentalized boxes and pad each section with soft paper.
- Do not use large boxes – filled with stones they become very heavy.
- Even hard crystals can be brittle. Avoid knocking them together.
- Natural terminations (ends) of crystals can chip easily, so store and handle with care.
- Avoid exposing crystals to very strong sunlight or sudden changes of temperature – colour can fade and fractures appear.
- Avoid placing shaped spheres in direct sunlight. They can focus the sun's rays and cause fire.

Cleansing and charging

New stones, provided they are not water-soluble or too fragile, benefit from washing in warm, soapy water to remove dust and grime. But more importantly, all your stones should be energetically cleansed before and after each use. This is because a crystal functions by absorbing energy from its surroundings into its internal structure, where imbalances are corrected. If it is not cleansed, the crystal will become overloaded, which reduces its effectiveness and may cause it to transfer imbalances into the aura. A "clean" crystal feels lighter and looks brighter than an energetically "tired" stone.

Crystal cleansing methods

Running water and sunlight Hold the stone under running water and visualize imbalances flowing away. Place in sunlight to dry. (Not suitable for water-soluble crystals, fragile crystals or crystals that fade in sunlight.)

Sound Strike or ring a singing bowl, bell, cymbal or tuning fork near the crystal for efficient energetic cleansing.

Incense smoke Incense has always been used to cleanse and consecrate. Burn some incense and pass the stone several times through the smoke.

Salt Place the crystal in a bowl and cover with dry sea salt. Leave to stand overnight. Avoid placing crystals in salt water as it corrodes the surfaces.

Breath Exhale sharply over the crystal and imagine imbalances being blown away. Repeat several times.

Crystal clusters Place the stone on any larger crystal cluster to restore natural levels of energy.

Essence sprays Spray the stone with a cleansing flower and gem essence.

Crystal healing techniques

One of the most effective ways of working with crystals is using layouts, which involves placing crystals in special patterns on and around the body to access and unblock the key paths of energy flow. You can also harness the power of crystals with pendulums and wands, as well as by integrating them into other therapies and traditions, such as reiki, colour therapy and feng shui.

For much crystal healing you can work alone or with another person. You will need to find a suitable space in which to work and spend a few minutes preparing for the experience. Crystal healing is cumulative. Regular sessions of five to ten minutes once or twice a week are more beneficial than long complicated processes practised less often. In this way your body will release stresses at its own pace as it learns to maintain equilibrium for itself.

Preparing the healing space

Although crystals can themselves create an atmosphere of calm and order, healing will be more effective if you work in a quiet, comfortable space.

- Find a room where you can lie down flat, with your legs outstretched.
- Reduce or remove distractions such as the telephone, television, strong light or loud music.
- Loosen tight clothing, remove watches, spectacles, shoes and jewelry. This will help the body to relax and enable energy to flow more easily around the body.
- When working with another person, keep everyday conversation to a minimum, but be attentive and open to sharing experiences.

Centring, grounding and protecting

All participants in a healing session should first centre themselves by focusing and quietening the body. This helps to avoid confusion, emotional upset, anxiety or distractions. To become centred, you can do any of the following:

- Hold or place crystals anywhere on the body's midline.
- Contemplate a large crystal held in both hands.

- Gaze at a crystal placed at a comfortable distance.
- Hold a crystal while imagining your breath moving through it.

You should also ground yourself (or -selves) before a healing session to anchor your energies and release stress. After a session, grounding crystals will discharge excess energy and integrate healing. Place one or two grounding stones – coloured black, metallic, brown or red – in one of these areas:

- on the front of the hips
- close to the base of the spine
- between the thighs
- between or on the knees

- between or below the feet
- 1m (3ft) below the feet
- pointing downward, along the length of the body

While crystals are in place during a healing session, relax with your eyes closed. Any sensations – such as tingling, a feeling of lightness or heaviness, energy surges or shifts of perception – are simply an indication that energies are balancing. If you experience any discomfort, remove all stones except grounding crystals. This will quickly restore calm.

Using layouts

Crystal layouts are set patterns of stones placed on and around the body. Each layout is designed to balance our vital energy in a particular way. Certain layouts, known as nets, call for us to lie upon a cloth of a particular colour in order to amplify the effect of the crystals.

Although most layouts are best carried out lying down, they may also, if necessary, be performed sitting up with stones taped lightly in place. If you are working alone, make sure that the crystals you need are within easy reach before you begin. You don't have to spend long with the crystals in position – five to ten minutes is ample time to feel the benefits. In most cases, you can use either crystal points or tumbled stones. However, some layouts do require crystal points, which direct energy more precisely than tumbled stones.

The diagram, below, shows an example of a layout to release stress, with tiger's eye on the sacral chakra, grounding stones by the feet, and rose quartz on the heart surrounded by four clear quartz points facing outward.

Wearing crystals

Gemstone amulets and talismans have long been worn or carried for protection. While wearing a crystal cannot in itself change your circumstances, a crystal that supports certain attitudes or behaviours can help *you* to change your circumstances for the better.

Grounding crystals – black, metallic, brown or red stones – strengthen the root chakra, which influences practicality and survival instincts. Multicoloured stones, such as labradorite or rainbow obsidian, are excellent at preventing other people from draining your energy. Pink stones, such as kunzite or morganite, emphasize calm, empathy and understanding, so can reduce the risk of situations escalating out of your control.

Sometimes, however, it can be more effective to strengthen your weaknesses. Consider the parts of your body that are vulnerable to pain or illness. Also think about situations that you find difficult mentally or emotionally. This may highlight certain chakras that need your attention. Look on page 15 for the colours of the stones that you could wear in times of stress to help you to maintain your energy.

All stones that you wear or carry should be cleansed regularly to remove the imbalances that they have picked up from you and your environment. Put jewelry on a small cluster of quartz or amethyst overnight to recharge it.

Wearing or carrying lots of different stones at the same time can be counter-productive, tiring and confusing the body. Alternate the stones you wear from day to day. Every so often take a rest from wearing or carrying crystals. This will help to prevent you from becoming dependent on their influences and will give you a better appreciation of their healing powers.

Crystal pendulums and wands

It's helpful to work with another person to get the most out of pendulums and wands. They are both effective tools for clearing imbalances from the aura.

Using pendulums

You can use a ready-made crystal pendulum (see right) or you can make one with a metal spiral attached to a length of thread – this has the advantage of allowing you to change the stones. Clear quartz, amethyst, smoky quartz and citrine are good stones to use.

- The person to be treated should lie down with his or her legs and arms out straight and relaxed.
- Hold the pendulum just above the area between and beneath his or her feet.
- Set the pendulum into a neutral swing (forward and backward in line with your forearm). From this starting point, move the swinging pendulum slowly and smoothly up the central line of the body all the way to the top of the head.
- Whenever the pendulum reaches an area of imbalance that it can remove, it will move differently while it disperses the imbalance, before regaining its neutral swing.
- Repeat the process, moving along each side of the body from below the feet to the top of the head.
- Repeat the process again, moving the pendulum a little way away from each side of the body, so you have passed the pendulum along five lines in all.

Using wands

Crystal wands are shaped with one rounded end and one pointed end (see below). The rounded end can be put directly on the skin. A crystal wand works in a similar way to a pendulum – the crystal moving through the aura helps to remove imbalances – but here the healer is more involved in the quality of movement. The technique described below is a good way to develop your intuitive skills, but it's important not to jump to conclusions regarding the cause or nature of any imbalances you sense. Even if you don't sense any imbalances, the technique will have a beneficial cleansing and calming effect.

- Ask the other person to lie down with his or her legs and arms out straight and relaxed.
- Hold the wand vertically between your thumb and fingertips with the point upward.
- Beginning just below the other person's feet, move your hand and arm anti-clockwise in small circular movements, while visualizing any energy imbalances being drawn out and neutralized through the crystal.
- Gradually move up the body in this way, spending as long as feels right on each area. In certain places you may need to change the speed, direction and pattern of movement. Allow your arm to move in whatever way it likes. Move on when you feel ready.
- When you reach the top of the head, turn the wand so that it is pointing down at the body of the person being treated. Repeat the process moving down the length of the body using clockwise movements of hand and wand.

Crystals and feng shui

There are many schools and traditions in the Chinese art of feng shui, but all have the same purpose: to bring balance and harmony to any environment – from a whole town or city to an individual house or room.

Feng shui works on the principle of enhancing the flow of *ch'i* (life-energy) – similar to the Indian *prana* (see page 14). One common system of feng shui seeks to balance the macrocosmic, universal patterns of direction and element (earth, fire, water, wood and metal) and the microcosmic, individual patterns of your living or working space. A room or building is divided using a set template, the *ba gua*, in which each of the sectors corresponds to a colour and an area of life. The grid, right, is a simplified version of the *ba gua*.

Wealth	Fame	Marriage or relationship
Family	Health	Children
Knowledge	Career	Helpful people and travel

Align this side with the main doorway

Where there is blocked or negative *ch'i* in a particular aspect of your life, feng shui recommends making changes of colour, structure or shape to restore harmony. Energy can be boosted by introducing an appropriate object or symbol, such as a plant or picture – or a crystal, especially a brilliant or faceted variety that can enliven dull areas by bringing light. If you want to enhance – or solve difficulties in – a particular area of your life, use the *ba gua* to place a crystal of the appropriate colour in the corresponding area of your home or a particular room.

Creating a sacred space

Many religions, such as Hinduism and Islam, and indigenous cultures, such as those of Africa and North America, share the tradition of setting aside a part of the home as a special place – a space dedicated to ancestors, household spirits, teachers and spiritual guides or simply for quiet contemplation.

Our homes are reflections of the way we conduct our lives, so if we don't have a quiet space, even if it's only a window-ledge or a shelf, then we're unlikely to take time to re-assess our actions, to take stock, to gain perspective and to let go of what we no longer need – all of which make a valuable contribution to our sense of well-being.

A sacred space doesn't need to have any religious images. The most famous Zen Buddhist gardens in Japan are simply weather-worn rocks placed on sand. A space becomes sacred when it allows us to move beyond our normal, everyday experiences, to gain insight and perspective. The space can be used to focus on healing or attaining a specific goal in life. If you decide to create your own sacred space, choose objects that reflect your intention. By spending a moment or two each day relaxing in that space we encourage the desired outcome by quietening our minds and focusing our actions.

A crystal or a group of crystals can make an effective meditative focus for a quiet space. It will also enliven the atmosphere and keep the space energetically clean. Choose a large, single crystal or a crystal cluster that you find fascinating and restful – one that you feel can hold your attention as you explore its many aspects. Spend a few moments a day simply gazing at the crystal, or crystals, in a relaxed way to reflect the attention you are giving to your own well-being.

Making crystal essences and waters

The energetic signature of a stone can be imprinted on water molecules. With water you can introduce the healing properties of a crystal to the body directly. Essences and waters enable the body to balance itself at its own pace.

Gem water

- Place a cleansed crystal or gemstone in a plain glass tumbler or pitcher. Fill with fresh spring water, cover and leave overnight.

- In the morning remove the crystal, and sip the water throughout the day.
- The water will keep in a refrigerator for a few days.

Gem essence

- Cover a cleansed crystal in spring water in a clear glass bowl and place in full sunlight for two to three hours.
- Pour the charged water into a bottle with an equal amount of vodka containing at least 40% alcohol (US 80 proof) to act as a preservative. Cork or cover the bottle.
- Pour a small amount of the essence into a smaller bottle with a dropper.

- Highly versatile, gem essences can be dropped onto the tongue, rubbed into pulse points or chakra points, sprayed into the aura, put in bath water, or sprayed around a room.
- Take at least three drops, twice a day – more if required.
- The dropper bottle can be refilled from the stock as required. Keep the stock bottle in a cool, dark place.

Take care! Only use sufficiently hard, non-water soluble, non-toxic stones to make gem waters and essences. Members of the quartz family are ideal. If in doubt, check a mineralogical textbook. *Never use*: realgar, orpiment, stibnite, galena or other compounds of mercury, arsenic or lead.

Crystals with other therapies

As well as being a healing tool in their own right, crystals can be used to amplify other therapies.

Hands-on healing, spiritual healing and reiki

In these techniques a healer draws on divine or universal energy and channels it to the patient. A healer may direct this energy through a crystal held in the hand and place a grounding stone beneath the feet to ensure that healing is clear, balanced and can be easily integrated.

Reflexology

This therapy works on the points and areas on the hands and feet that correspond to the organs of the body. A reflexologist encourages healing by stimulating reflex points with massage and other manipulation skills. A crystal wand or sphere can be used instead of fingers. Crystals placed by the feet of the patient, before and after reflexology, stabilize and deepen the healing effects.

Colour therapy

Colour, by itself, is a very strong healing tool. Its effects can be softened with the use of crystals, which combine the power of colour with the coherent, modifying influence of crystal structure. A healer can gain great insight into the subtle energies at play in a situation from the patient's choice of crystal. Our favourite colours support us and lift our energies. The colours we avoid represent the energies we are blocking, which may be more beneficial to us – so, paradoxically, the colours we dislike are often the ones we actually need.

A quick reference crystal finder

In the Guide to Crystals (see pages 46–275), we will explore the properties of individual crystals. Here, we suggest crystals to treat the major parts of the body, body systems and aspects of the mind, body and emotions. Use the list, below, to home in on the most appropriate crystals for you; then refer to the entries for those stones in the Guide to learn how to use them.

• **Head and eyes** azurite, amethyst, fluorite, sugilite, diamond

• **Ears, nose, throat and neck** amazonite, aquamarine, angelite, blue lace agate, blue quartz, celestite, tanzanite, lapis lazuli, iolite, kunzite, Preseli bluestone

• **Chest and lungs** moss agate, chrysocolla, larimar, turquoise

• **Heart** ruby, prase/actinolite quartz, aventurine, seraphinite, dioptase, emerald, coral, labradorite, unakite, watermelon tourmaline, verdelite

• **Digestion** bronzite, citrine, pyrite, sulphur, prehnite, purpurite, zircon

• **Large intestine, kidneys, bladder** obsidian, peridot, jade, chalcopyrite, ulexite

• **Reproductive organs** silver, carnelian, morganite, rhodocrosite, milky quartz, moonstone, selenite, opal, pearl, rubellite

- **Circulation** haematite, iron quartz, bloodstone, sandstone

- **Nervous system** citrine, silver, gold, sapphire, rutilated quartz

- **Immune system** turquoise, aquamarine, larimar, clear quartz, tourmaline

- **Bones and muscles** flint, black tourmaline, magnetite/lodestone, spinel, rutilated quartz, apatite, tourmaline quartz

- **Boosting physical energy levels** jet, smoky quartz, limonite, tiger's eye, garnet, jasper, amber, gold, diopside, serpentine, mookaite, basalt

- **Enlivening the mind** meteorite, tektite, petrified wood, vesuvianite, zincite, moldavite, sphene, apophyllite, danburite

- **Quietening the mind** staurolite, calcite, chrysoprase

- **Reducing stress** onyx, aragonite, copper, malachite, thulite, granite

- **Stabilizing the emotions** halite, sunstone, topaz, heliodor, epidote, kyanite, smithsonite, dumortierite, sapphire, sodalite, charoite, lepidolite, rhodonite, rose quartz, pietersite, limestone, chalk, marble, zeolite, clear quartz

Crystal layouts

In the previous section we introduced the principles of crystal layouts (see page 28) – a fundamental way of harnessing the healing power of crystals by placing stones on and around the body. Here, we suggest a range of the most useful layouts to incorporate into your crystal-healing routine.

Simple daily layouts

To get the most out of crystals, you need to use them regularly. If you take a few minutes each day to carry out the following two basic layouts, you will help to prevent the build-up of imbalances and stresses.

Clarifying layout

This layout clarifies awareness and releases blocked energy. It is useful when you are feeling unsettled for no apparent reason. With the release of stress, sensations may seem to increase before they subside.

- Place smoky quartz below the feet.
- Place clear quartz above the head.
- Place either turquoise (to strengthen vitality) or lapis lazuli (to release deep stress) in the centre of the forehead.
- Stay in layout for five to ten minutes.

Balancing layout

This layout provides a good everyday balance and restores a natural flow of energy.

- Place smoky quartz by the feet.
- Place clear quartz above the head.
- Place rose quartz on the centre of the chest.
- Stay in layout for five to ten minutes.

Rainbow chakra balance

To recharge and balance the body's energy centres, the chakras (see pages 14–15), lie down and place the appropriate coloured crystals on and around the body as described below. If you have a selection of crystals that could be used for each colour, choose by intuition.

- First, place a grounding stone – black, metallic, brown or dark red – between the feet.
- Then put a red stone close to the root chakra – near the base of the spine or between the knees.
- Next, position an orange stone at the sacral chakra, just below the navel.
- Put a yellow stone at the solar plexus chakra, below the ribs.
- Place a green stone at the heart chakra, in the centre of the chest.

- Put a light blue stone at the throat chakra, at the base of the throat.
- Then position an indigo (deep blue) stone at the brow chakra, in the centre of the forehead.
- Finally, place a violet or clear stone at the crown chakra, just above the top of the head.
- Lie and relax in this position for five minutes, then remove the stones from top to toe, leaving the grounding stone until last.

Variation You can also choose crystals to balance the chakras using intuition alone and taking no account of colour. Sit with your collection of crystals in front of you, close your eyes, quieten your thoughts, then focus on the root chakra. After a minute, look at your crystal collection and pick out the first stone you notice. This will balance the root chakra. Repeat this process for each of the other six chakras. Then take the seven stones you have selected and follow the steps given above for the rainbow chakra balance.

Easing the body

Given the right environment and support, the body itself can treat many physical ailments. The following two layouts quieten and relax every level of our being, allowing the flow of healing.

Seal of Solomon

Use only clear quartz crystals in this layout for general physical healing. You can also put any other kind of stone in the centre of the layout to focus the healing qualities of that stone. For a localized pain, create a smaller version of the layout around the affected area. A green or blue stone placed on an inflamed area will calm and cool.

- Place six clear quartz crystals, points outward, around the body: one above the head; one by each upper arm; one by each thigh; and one centrally beneath the feet.
- After a few minutes, or when it feels right, turn the points inward. The process will be complete after a further two or three minutes.

Figure of eight

This pattern revives sluggish energies when we feel agitated. Like the Seal of Solomon, it can be localized to any area of the body that needs bringing back to health.

- Place eight or more smoky quartz points or black tourmaline crystals around the body in a figure of eight pattern, with all the points in the same direction.

- Begin with a stone at the left shoulder, followed by one each in the centre of the chest, by the right hip, right calf, centrally beneath the feet, by the left calf, left hip and at the right shoulder.
- If you have more stones, use them to fill gaps in the shape.
- Stay within the layout until you feel settled.

Calming the mind, lifting the spirits

Any experience that upsets our equilibrium gets "stuck" in our system in the form of stress, which prevents the natural flow of energy. The layouts, below, help to unblock stresses from all levels of our being.

Time out
When you feel anxious, this layout will help to quieten your mind.
- Place a lapis lazuli crystal, or another dark blue stone, in the centre of the forehead and surround the blue stone with three small fluorite pieces in an upward-pointing triangle.
- Place another piece of fluorite just above the top of the head.
- Stay in layout for five to ten minutes.

Joy giver
This layout will bring you joy if you are feeling low.
- Lie down on a yellow cloth, ideally with your head pointing north.
- Place six quartz points around the body as follows: one above the head; one at each shoulder; one on the solar plexus chakra; and one beside the top of each foot.
- Stay in layout for five to ten minutes.

Lighthouse
If you feel isolated or disconnected, this layout will give you spiritual guidance.
- Lie down on a pink cloth.
- Place a black tourmaline centrally below the feet.
- Place a green tourmaline on the cloth either side of the body, level with the heart.
- Place a blue tourmaline above the top of the head, with a pink tourmaline above the blue stone.
- Stay in layout for five to ten minutes.

Crystal meditations

It's hard to resist gazing at an attractive stone, and crystals are an ideal focus for meditation. As with layouts, you should find a quiet place to meditate, and it's good to prepare by centring and grounding yourself (see page 27).

Chakra meditation

This is both a healing exercise for balancing the chakras and a meditation.

- Assemble the black, metallic, brown and red stones of the root chakra.
- Pick the one that most attracts you. Hold it in both hands for a few minutes. Let its motivating and stabilizing energy flow through you.
- Repeat this process for each of your other chakras in order. Choose:
  - an orange stone for your sacral chakra and absorb from it the energies of creativity, flow, healing and pleasure
  - a yellow or gold stone for your solar plexus chakra and absorb sustenance, organization and clarity
  - a green or pink stone for your heart chakra and absorb the qualities of calm, balance, personal direction and space
  - a turquoise or light blue stone for your throat chakra and absorb peace, communication, creativity and personal expression
  - a dark blue stone for your brow chakra and absorb intuition, insight and broad perspective
  - a violet, white or clear stone for your crown chakra and absorb vitality, and integration with the energies of the universe
- Finish by selecting one stone from all the colours in front of you. Hold it for a moment and sense it restoring equilibrium to all your energies.

Pink heart star meditation

The colour pink is excellent for reducing fear, aggression, anger, irritation and misunderstanding. Pink stones protect us from all forms of negativity.

- Search through your collection of crystals for five pink stones that you find particularly appealing.
- Sit on the ground and place four of the stones around you: to the front, back, left and right.
- Hold the fifth stone in your hands.
- Focus your attention on your heart chakra. Close your eyes and imagine a spark of bright pink light coming to life in the centre of your chest.
- See that pink spark begin to shine like a star, increasing in strength and radiance, filling your body with a strong pink glow.
- As the glow spreads outward, visualize it touching the stone that you are holding.
- The glow continues to spread until it reaches the stones around you.
- Imagine that each stone also begins to radiate pink light from a star at its centre.

- Maintaining the pink heart star immediately around you, see the pink light continue to radiate out into your surroundings.
- Imagine the light touching the hearts of people you know and sparking stars in them, which in turn radiate out to touch the hearts of the people they know.
- Continue this meditation as long as you wish, focusing particularly on directing the light toward anyone with whom you are having difficulties.
- To end the meditation, simply allow the light and your attention to return to the stones around you, then to the one in your hand, and finally to your heart.
- Allow the colour gradually to fade away, then open your eyes.

Meeting crystal spirits

This meditation allows you to meet the spirit of a crystal. The first few times you try this, you may not be able to meet the spirit in any definite form. But after a few explorations with the same crystal, set out with the intention of meeting the spirit in a form with which you can easily communicate, such as a person or animal. The spirit will help you to understand how you might benefit from working with that crystal. The clearer your intentions when entering the meditation, the easier it will be to understand your experiences.

- Hold the crystal you wish to explore in your hands or place it upon the chakra that corresponds to the colour of the crystal.
- Imagine the mouth of a cave in front of you. Take time to build the image clearly.
- Picture yourself tentatively moving forward into the dim light of the cave. Move slowly, exploring your impressions.
- Move further forward and see in front of you a downward-leading path going deeper into the Earth.
- Follow the path. You will eventually arrive at a closed doorway.
- Imagine a picture of the crystal or its name on the door. Focus on your intention to meet the crystal spirit and wait for the door to open.
- When the door opens, step through and gather your impressions. Take time to explore – this is the point at which you are to meet the crystal spirit. Everything you experience beyond the doorway is an expression of the stone's nature.
- When it's time to return, say "thank you" and will yourself back to the doorway.
- Walk through the door and close it behind you. Make your way back up the path and out of the cave.
- Take time to consider and record your experiences before resuming everyday activities.

Exploring crystals

Crystals are stable and have constant healing properties and a constant level of energy. In contrast, humans are unique and continually readjusting to their internal and external environment. If a crystal feels energizing one day but calming the next, it is not the crystal that has changed, but the needs of your body. This meditation will help you to understand how the effects of a crystal can alter.

If you repeat this procedure several times, even if you are not aware of vastly different effects, you will have become more attuned to the range of that crystal's properties and so better equipped to use it appropriately. Using this technique regularly will also strengthen your awareness and intuitive skills.

- Make sure that the crystal you want to investigate has been cleansed before you begin (see page 25).
- Take a few minutes to look at your chosen stone closely.
- Hold it in both hands, close to your solar plexus (just below your ribs) for a minute or two. Imagine that your breath is passing through the stone. This helps you to tune into its energy.
- Now place the crystal a little in front of you so that you can gaze at it in a relaxed way.
- Consider how you feel physically. Note any parts of your body that draw your attention.
- Now consider your emotions. What moods are you feeling? What is the quality of your thoughts?
- Pick up and hold the crystal again.
- Imagine breathing in and out through the stone for a minute or so.
- Consider how you feel now. Where does your attention go to in the body? Has your mood altered? Are your thoughts different?
- Finish by grounding yourself.

Part Two:
A Guide to Crystals

In this part of the book we bring you an extensive directory of more than 100 crystals that should be easy to acquire and that are safe and practical to use in everyday crystal healing. Because colour is such an important aspect of the way that crystals function in healing, we have grouped stones according to the colour in which they are most commonly found. We explore the magical and healing properties of each crystal, and suggest practical ways to harness its unique powers. This will help you to learn which crystals will be of benefit to you at any given time. You will also find key geological information, such as colour variations, lustre, hardness and lattice system, as well as tips on how to care for and identify the stone – particularly how to distinguish it from similar-looking crystals.

Black tourmaline

$(Na(Mg,Fe,Li,Mn,Al)_3Al_6(BO_3)Si_6.O_{18}(OH,F)_4)$

Black tourmaline, also known as schorl, is the most widespread variety of the very common mineral, tourmaline. It gets its black colour from the amount of iron it contains. In healing terms, black tourmaline has a strong, clear energy, which makes it particularly effective. From dispelling negative energies in the home to releasing pulled muscles and reducing jet lag, it is invaluable. When heated, black tourmaline becomes negatively charged at one end and positively charged at the other. This hints at one of the stone's key healing properties: the ability to help us to balance extremes.

Colour: black	
Lustre: vitreous	
Hardness: 7–7.5	
System: trigonal	

Identification and care

- Single crystals are easy to identify by their curved triangular cross-section. Terminations are a low, even dome.
- If thin crystals are held up to the light, they often take on a brown or green tinge. The sides of the crystals have parallel striations.

A single black tourmaline crystal

Magic

- Deflects any potential harm, so is one of the best crystals to protect from negative forces
- Has strong links to the energy of the Earth

Healing functions

- Realigns the structures of the body
- Has a stabilizing and anchoring effect as it is closely related to the root chakra
- Soothes and settles bones and muscles
- Focuses awareness in the present
- Strengthens our grip on reality
- Grounds and protects us

Practical ideas

- Wear tourmaline earrings or fix small pieces of the stone behind the ears with sticking plasters to ease tinnitus and earache.
- Carry tourmaline when travelling to reduce tiredness, jet lag and disorientation.
- Place under a bed or mattress to achieve restful sleep, particularly in unfamiliar places.

Keywords

Grounding

Protecting

Realigning

Similar stones

Epidote crystals are identical in structure to black tourmaline but they are usually dark brown-green.

Onyx, when tumbled, is hard to tell from black tourmaline, unless it has lighter banding.

Tumbled **obsidian** looks very similar to tumbled black tourmaline, but is more translucent.

Obsidian (complex silicate with inclusions)

Formed by rapidly cooling lava, obsidian is found mainly in volcanic regions. Its black colouring comes from its iron content, and it is more like glass in structure than crystal. This can sometimes make it feel "dead" in comparison with other stones. A little more attention, however, reveals a busy background "noise" that can help to break down outworn patterns in our lives.

Colour: black, greenish black, grey or red-brown	
Lustre: vitreous	
Hardness: 6	
System: amorphous	

Identification and care

- Obsidian can vary from a pale, smoky, translucent variety called "Apache tears" to completely dense, opaque, black glass.
- Snowflake obsidian contains grey or white flecks, which are crystals of feldspar, mica, and other similar minerals.
- Mahogany obsidian has red-brown streaks.

A rough piece of obsidian showing razor-sharp edges

- The iridescence of rainbow obsidian is caused by tiny crystals defracting light.
- Unpolished or untumbled obsidian often has razor-sharp edges, so must be handled with care.

Magic
- Traditionally polished to make scrying tools
- A smooth disk or polished sphere can be used as a mirror to aid clairvoyance
- Reveals what is hidden or lost

Healing functions
- As a stone that emerges from the ground with dramatic force, is able to bring hidden issues, emotions and traumas to the surface
- Rebalances the digestive system
- Grounds and protects us

Practical ideas
- Space five obsidian around the body in a pentagon shape, with the point above the head to encourage change. Lie on a red or black cloth to deepen the experience.

Keywords

Revealing

Cleansing

Transforming

Similar stones

Black onyx is very difficult to distinguish from obsidian. Obsidian, however, can sometimes be more translucent.

Tumbled black tourmaline often feels heavier than obsidian.

Tumbled smoky quartz looks almost identical to "Apache tears".

51

Onyx (SiO₂)

Onyx is a variety of agate comprising bands of quartz, chalcedony and opal that have been laid down in coloured layers within rock cavities. The stone's name means "fingernail" in Greek – its thin white bands often resemble the curved edge of a fingernail. Onyx could not be described as a fun-loving, extrovert stone. Its stark, rather sombre black and white bands are often impressive but restrained. However, sometimes we need to step back and look dispassionately at our situation in life, and onyx is ideal for this as it can help to reveal all viewpoints, all polarities, all apparent conflicts. In contrast, sardonyx, an orange version of onyx, has warmer, more engaging associations.

Colour: black and white bands, sometimes just black

Lustre: vitreous, silky or dull

Hardness: 7

System: trigonal

Identification and care

- Onyx is easily identifiable by the striking contrast of its colours (black and white).
- Slices are hard but brittle.

Magic

- Has a calming effect, which can be used to balance excitability and a quick temper

Similar stones

All tumbled black stones, such as **black tourmaline**, **obsidian** and **smoky quartz**, can be difficult to distinguish from onyx unless it has banding.

- Increases introspection, but can also lead to inertia and withdrawal from everyday activities in some people
- Can make the user or wearer less noticeable to others if desired

Keywords

Calming

Distancing

Quietening

Healing functions

- Grounds and concentrates energy while helping to cleanse and purify
- Alleviates problems in an orderly way
- Quietens powerful emotional states
- Encourages stillness and introspection
- Reveals underlying causes of situations

Practical ideas

- Gaze at a banded onyx sample to explore hidden aspects of a situation.
- Combine onyx with an activating stone, such as jasper or garnet, to achieve insight and reassessment without alienating yourself from other people.

Small pieces of tumbled onyx　53

Magnetite and lodestone (Fe^{2+}Fe$_2^{3+}$O$_4$)

In contrast to other iron oxides (like haematite), which often contain impurities, magnetite (known as lodestone when it has become magnetized) is pure. When the Earth formed, the heavier metals, such as iron, were pulled by gravity into the planet's core, where the Earth's magnetic field is generated. Magnetite can therefore help us to remain in harmony with the planet. The earliest compasses incorporated pieces of magnetite, and in healing, too, it helps us to avoid loss of direction. And no other stone can so speedily neutralize the environmental pollution of modern urban living.

Colour: black

Lustre: metallic, dull

Hardness: 5.5–6.5

System: cubic

Identification and care

- Single, octahedral crystals of magnetite are common.
- Lodestone is magnetically polarized, so often has iron filings attached to it.
- Tumbled lodestone is shiny with an uneven surface.
- Magnetite tends to be more grainy, with angular fractures.

A raw piece
of lodestone

attached iron filings

Magic
- Helps us to find the right path
- Turns adverse situations around

Healing functions
- Aligns the chakras and subtle bodies
- Releases stress, grounds and energizes
- Realigns the body to the electromagnetic field of the planet

Practical ideas
- Rotate a pendulum of lodestone through a person's energy field to remove electro-magnetic pollution from the aura. This is useful for those who feel drained of energy after using a computer or watching television.

Similar stones

Haematite surfaces are more reflective.

Chromite is an oxide of chromium plus magnesium. It is less magnetic and leaves a brown streak (magnetite and lodestone make a black streak).

55

Jet (C)

Jet is the waterlogged wood of prehistoric monkey puzzle trees, compacted by pressure under the sea and washed ashore when storms disturb its underwater beds. It was popularized in the 19th century by Queen Victoria, who wore jet from Whitby, Yorkshire, in mourning for her husband, Prince Albert. Many dark, grounding stones originate deep within the planet as volcanic materials and so have a fiery, seething quality. Jet, on the other hand, is calm, cool and balanced. It tends to be quite a gentle way to "return to oneself". Working with jet often feels somewhat like sitting beneath the ancient branches of a protective tree.

Identification and care

- Jet is extremely light for its size, and can, at times, resemble plastic.
- A piece of jet generates static electricity when rubbed.
- Completely untreated jet has dull, satin-like surfaces.
- Annual rings, like those on a cut tree trunk, may be visible in natural samples of jet.

Colour:	black, dark brown
Lustre:	vitreous
Hardness:	2.5
System:	amorphous

Similar stones

Onyx is much heavier than jet.

Black tourmaline, when tumbled, is much heavier and more glassy than jet.

"Paris jet" is actually black glass.

Magic

- Originating in Earth's ancient forests, jet allows us to see beyond the present moment and perceive hidden things.
- Highly polished jet can be used as a scrying mirror for divination.

Healing functions

- Activates energy in the spine
- Lifts feelings of depression
- Calms the mind
- Engenders feelings of being safe, comforted and protected

Practical ideas

- If you feel overburdened by worries or are over-analyzing your situation, place jet below your waist (in a pocket or on the body) to balance energy through the whole body.

Keywords
Stability
Calm
Perspective

Natural jet (one face polished)

Meteorite (composition varies)

Many precious stones, such as rock crystal (clear quartz), were thought to have descended from heaven. Meteorites are actually seen to fall from the sky, so have always been held in special regard. Some people believe that they are the remains of a large planet, similar in composition to the Earth, which once existed in our solar system, in the asteroid belt between Mars and Jupiter. There are two main types of meteorite but iron meteorites are the more common. These are often mixed with nickel and are very heavy. There are also stone meteorites, similar in structure to igneous rocks, often containing olivine crystals. Rarer meteorites, known as pallasites, consist of both stone and metal.

Colour: light grey, grey-green or black	
Lustre: metallic, dull	
Hardness: variable	
System: variable	

Identification and care

- Meteorite can be found anywhere in the world.
- It often has a smooth, undulating surface and is very heavy for its size.

Natural pieces of meteorite from Russia

- Meteorite with a high iron content is quite brittle.
- When a meteorite is cut open and etched with acid, its internal pattern of alloys is clearly visible.

Magic
- Has the potential to open up new levels of awareness
- Traditionally, highly prized and forged into magical jewelry and weapons

Healing functions
- Boosts our strength and resilience
- Activates creative mental skills, such as intuition, imagination and inspiration
- Broadens perception

Practical ideas
- Hold meteorite for inventive solutions to problems. It often feels energizing and "busy" and will activate inspiration and imagination, as well as amplifying the processes begun by other stones.

Keywords

Perception

New horizons

Awakening

Similar stones

Olivine-rich rocks look very similar to rocky meteorites.

Iron and nickel ores lack the molten-looking exterior of meteorite, as well as the interior crystal etching.

Flint (SiO₂)

Flint is a variety of quartz mixed with chalcedony, which was formed by sea fungi and silica-rich shells laid down in ancient oceans. When broken, flint has very sharp edges, which made it useful to our ancestors for tools and weapons. It was also used for making fires as striking flint creates sparks. Often ignored as a plain, unappealing stone, flint represents humankind's long and ingenious association with the technologies and the magic of stone, and reminds us that the beauty of an object is not only in its appearance, but in its potential uses. Natural flint resembles bone and so became associated with ancestors and Earth spirits. No stone can give us a deeper feeling for our history.

Identification and care

- Flint can be collected from the countryside in limestone areas, and is particularly common on the seashore.
- It is usually found in irregular, rounded nodules, the surfaces of which become powdery-white on exposure to air.
- Edges are translucent and can be razor-sharp.

Colour: blue-grey, grey or black

Lustre: dull to vitreous

Hardness: 7

System: triclinic

Similar stones

Botswana agate is often grey, like flint, but it has clear white banding.

Quartzite is whiter than flint.

Hornstone is very similar to flint, but is of inorganic origin and found in veins of ore.

- Flint has characteristic conchoidal
  (shell-shaped) fracture lines.

Keywords
Repair
Knowledge
Confidence

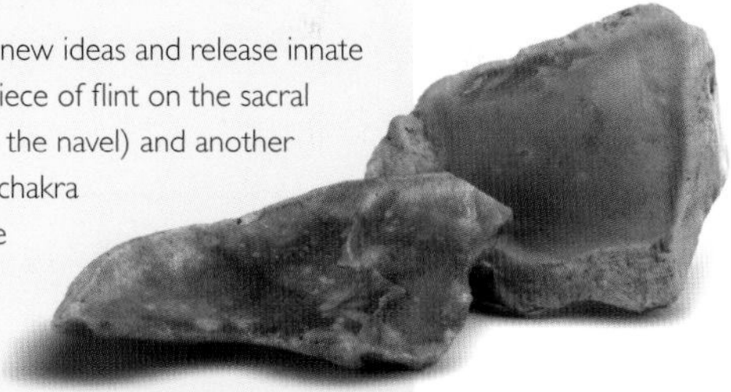

Magic
- Traditionally believed to divert harm and
  increase our knowledge of the Earth
- Protects us against nightmares

Healing functions
- Helps the body to absorb essential nutrients
- Accelerates the healing of broken bones
- Enables us to assimilate information
- Enlivens chakras and subtle bodies

Practical ideas
- To encourage new ideas and release innate
  skills, place a piece of flint on the sacral
  chakra (below the navel) and another
  on the throat chakra
  (just above the
  sternal notch).

Natural pieces of flint

61

Haematite (Fe₂O₃)

This commonly occurring mineral has a high iron content. Usually found in rounded, kidney-shaped lumps, it was dug out of the world's first mines, in Africa. It can be used to make a red pigment, which was once sought after as a sacred substance. Of all grounding stones, haematite is the most useful for bringing our awareness to the here and now. In a very small proportion of people, however, it has the opposite effect: in these cases, the reflective polished surface opens awareness to the hidden, magical worlds of spirit.

Colour: metallic grey-black or red

Lustre: metallic or dull

Hardness: 5–6

System: trigonal

Identification and care

- Tumbled stones are shiny, metallic and heavy for their size.
- A rough sample is dull.
- Broken edges turn red and powdery when exposed to the air.
- It is brittle, so shatters easily if dropped or struck.

A natural piece of haematite

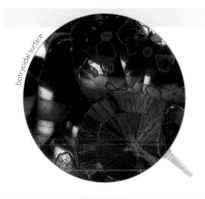

botryoidal surface

Magic

- Conveys power and invulnerability
- Protects against all harm

Healing functions

- Supports the blood and circulatory system
- Reduces inflammation
- Boosts self-esteem
- Grounds and anchors us
- Energizes all aspects of mind and body

Practical ideas

- Wear haematite jewelry to anchor spiritual activity in reality.
- Carry haematite to restore energy when you are feeling lethargic or "spaced out".

Similar stones

Tumbled **magnetite** is less shiny and mirror-like.

Marcasite, another shiny iron ore, is rarely seen in large pieces.

Silver (Ag)

Silver has been valued and used for many thousands of years. Today it can be found in its pure, native state, as well as in various compounds. It occurs in masses of silver wires, irregular shaped plates and sheets, as well as in the form of solid nuggets. Just as gold represents the sun, so silver epitomizes the moon – recalling the cool, contemplative, feminine light of the night sky and the romantic reflection of moonlight off water. Silver can be used to cool overactive energetic states, as it helps to create flow and movement. It is introspective and aloof in contrast to the powerfully extrovert display of solar gold.

Colour: silver-white that tarnishes to grey-black

Lustre: metallic

Hardness: 2.5–3

System: cubic

Identification and care

- The brilliant shine of the freshly revealed metal soon tarnishes to black in contact with air.
- Pure silver is very soft and is usually mixed into an alloy.

A sheet of native silver alloyed with bornite

64

- Silver is uncommon but worth looking out for – it is usually sold in thin plates or as nuggets of silver and nickel alloy.

Healing functions

- Eases the flow of energy in the body and clears it of impurities
- Eases difficulties with movement
- Alleviates nerve twitches
- Prevents infections
- Improves fertility
- Aids mental balance
- Encourages clear insight and intuition

Practical ideas

- To protect against infection or to soothe when you feel unwell, place a cleansed silver sample in a glass of drinking water for an hour or two. Remove the silver, dry it, and sip the water occasionally during the day.

Tumbled silver-nickel

Keywords

Easing

Serenity

Flow

Similar stones

Galena (lead sulphide) is as soft as silver, but much heavier. It is poisonous to the touch and not recommended for crystal healing.

Stibnite (antimony sulphide) has long striated metallic crystals that look like the "lead" in pencils. Like galena, it is potentially poisonous.

Also: **marcasite, graphite**

Smoky quartz (SiO₂)

Smoky quartz can be found where granite exists, often in mountainous regions or within volcanic rocks. The smoky-black colouration is thought to occur when clear quartz is subjected to radioactive decay in nearby minerals.

In healing, smoky quartz has a more gentle effect than clear quartz, which can be over-energizing for sensitive people. Smoky quartz has as many varieties and moods as shades of colour. It can feel peacefully radiant, absorbing, solid, earthy and mysterious. With its perfectly black crystal facets, it absorbs the attention in a quiet way that the reflective brightness of clear quartz can rarely achieve. Within it, one can sense the depths of space, the potential of new creation and the memory of ancient events.

Colour:	smoky brown to brown-black
Lustre:	vitreous
Hardness:	7
System:	triclinic

Identification and care

- Much smoky quartz in shops is artificially irradiated clear quartz but it can be used in the same way as natural smoky quartz.

An elestial smoky quartz

- Morion is a deep black variety.
- The yellow-brown variety found in Scotland is called Cairngorm.

Magic

- Concentrates energy (whereas clear quartz expands energy)
- Provides a focus for meditation

Healing functions

- Calms the mind and focuses its thoughts
- Gently grounds us

Practical ideas

Similar stones

- Hold clear quartz in one hand and smoky quartz in the other to balance "receiving" and "broadcasting" energies. Experiment to find the positions that are restful and those that are stimulating.

Citrine quartz (see left) resembles pale versions of smoky quartz.

"Apache tears", a variety of obsidian, is almost identical to smoky quartz.

- Place 12 smoky quartz crystals evenly around you, pointing away from the body, for a powerful healing and clearing effect. Lie on an orange cloth to deepen the experience.

Tektite (SiO$_2$)

Tektite can often resemble old forgotten boiled sweets (hard candy) or bits of dried tar. But its origins are mysterious – rock and soil flung together and melted by meteoric impacts millions of years ago. Tektites are named after the places where they are found. For example, australite is found in Australia, javanite in Java and moldavite (see pages 154–155) in central Europe. One of the most unusual is "Libyan glass", a translucent yellow-green tektite found strewn across the surface of the Libyan Sahara and thought to have been created by some cataclysmic event in the distant past.

Colour: various shades of green, brown, black	
Lustre: vitreous	
Hardness: 6–7	
System: amorphous	

Identification and care
- Surfaces tend to be irregular and pitted.
- Some tektites can be cut into gemstones.
- Tektites are rarely translucent except at thin edges, where they show brown.

Magic
- Amalgamates the energies of terrestrial and extra-terrestrial influences
- Defuses volatile situations

Similar stones

Meteorite has the same surface pitting, but tektite is lighter and more glassy.

Obsidian does not have surface pitting.

Volcanic "bombs" are very similar to tektite, but may look less weathered.

Furnace slag contains more impurities and looks less weathered.

Healing functions

- Encourages integration of disparate energies, with a practical focus
- Neutralizes over-emotional states
- Helps us to find solutions to problems
- Brings us back to the real world after flights of fancy

Practical ideas

- In any intense state of confusion, place a tektite on the area of the body where you feel the focus of the agitation. Combine with clear or smoky quartz – as grounding stones or in a Seal of Solomon layout (see page 40) – to rapidly balance the situation.
- Meditate with tektite, or place it under your pillow to encourage new perspectives, new ideas and new ways to express personal creativity.

Keywords
Containing
Amalgamating
Transforming

Rough pieces of tektite

69

Tiger's eye (SiO₂)

A member of the quartz family, tiger's eye consists of fibres of the common rock-forming mineral amphibole – usually the variety known as crocidolite – embedded in quartz in highly packed bands. The silky blue-black fibres of crocidolite give the quartz a blue-green sheen and produce a stone known as falcon's eye. If these inclusions oxidize, tiger's eye is created, with its characteristic brown and gold colour. Streaks of blue crocidolite may also still be present. If exposed to heat tiger's eye turns red, but most red tiger's eye that you can buy has been artificially heated. The fine lines within tiger's eye underscore one of its functions – that of a good energy shifter. Like all stones that resemble eyes when cut or polished, tiger's eye is a powerful protective amulet.

Colour: brown, gold
Lustre: vitreous
Hardness: 7
System: trigonal

Identification and care

- As light reflects off the fibres in the stone, the silky stripes of brown and gold seem to move and change.
- Round cabochons allow the maximum play of light and are often used in jewelry.

70

A piece of natural tiger's eye

- The sheen depends upon the angle of cut.
- The blue-green sheen of falcon's eye is often more subtle than that of tiger's eye.

Magic
- Protects against the "evil eye", demons and witchcraft
- Diverts unwanted energy
- Confuses your opposition

Healing functions
- Releases stuck or congested states, for example, in the digestion or in the mind at times when thought processes are confused
- Soothes, both physically and mentally
- Builds confidence
- Encourages contact with other people

Practical ideas
- To activate practical energy: place tiger's eye on the root, sacral and solar plexus chakras.
- To stimulate communication and the flow of ideas: place blue falcon's eye on the brow and throat chakras.

Keywords

Practicality

Sociability

Realism

Similar stones

Tiger iron consists of tiger's eye, red jasper and black haematite and has rippled, wavy bands of colour.

Rutilated quartz (see below) is more yellow and has a less regular "sheen".

71

Vesuvianite $(Ca_{19}(Al,Mg,Fe)_{13}Si_{18}O_{68}(OH,F,O)_{10})$

Named after the volcano Mount Vesuvius in Italy, vesuvianite is another name for idocrase. It forms in metamorphic conditions where there is limestone or dolomite. Vesuvianite's clarity and variety of colours make it both a useful and attractive gemstone. One of its more appealing forms comprises a large central crystal surrounded by smaller radiating crystals. The 18th-century poet-scientist Johann Wolfgang von Goethe found it so beautiful that he wrote an epic poem about the stone.

Colour: brown, green, black-green, yellow (rarely blue, purple or colourless)

Lustre: vitreous, dull

Hardness: 6–7

System: tetragonal

Identification and care

- Vesuvianite commonly forms long prismatic crystals or short pyramidal crystals.
- Columnar crystals have a square cross-section.
- Idocrase occurs in a variety of colours, named for where they were found: egeran is brown; californite is green; vilyuyite is black-green; zantite is yellow; cyprine is pale blue.
- Its crystals are quite brittle. Handle with care.

Tumbled vesuvianite

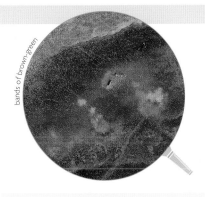

bands of brown-green

Magic

- Brings about transformation and creates illusions
- Can cause a change of appearance
- Suggests a new way of looking at situations

Healing functions

- Helps to bring resolution and stability
- Reduces confusion and clarifies complex situations

Practical ideas

- To find a solution to a difficult problem: place a vesuvianite stone at the throat chakra, one at the heart chakra, and hold one in each hand. Repeat if necessary.

Similar stones

Zircon crystals are shaped like pointed barrels.

Epidote is very hard to distinguish from vesuvianite.

Grossular garnet feels heavier than vesuvianite. Gem-quality crystals have a bright green translucence; more common examples have a pale opaque, dull sheen.

Topaz has characteristic chisel-shaped crystals.

73

Halite (NaCl)

Halite is the name given to rock salt. The halite crystals found today were formed during dry, hot periods of prehistory when ancient inland seas evaporated. Halite is often found alongside other mineral evaporates, such as gypsum. Mined since the earliest part of the Stone Age, salt has long been a valuable commodity. As well as being an essential element of the human diet (in moderation) and an effective preservative, halite crystals are a powerful healer. However, they are highly absorbent, so they must be handled and stored carefully.

Colour: pink, brown, white, yellow, blue

Lustre: vitreous

Hardness: 2

System: cubic

Identification and care

- Halite itself is transparent. Its colour comes from impurities of other minerals.
- Halite crystals are very soft. Store separately.
- When struck, halite will divide into smaller cubic crystals.
- Keep halite in dry conditions. It absorbs atmospheric water.
- Never clean halite in water – it will dissolve.

Natural halite crystals

Magic
- Protects from evil forces
- Deflects negative influences

Healing functions
- Keeps the nervous system working effectively
- Helps to maintain the critical salt balance in the body, where too little causes weakness and confusion and too much causes water-retention and muscle damage
- Balances the subtle systems of the body
- Helps to stabilize the emotions

Practical ideas
- Halite lamps, consisting of hollowed-out crystals containing a bulb or candle, emit a warm, calming light and clean the air of smoke, dust, mould and electrical pollution.
- To remove environmental pollution: place bowls of salt or small pieces of halite around the home. Throw away the salt when the process is complete.

Keywords
Absorbing

Protecting

Stabilizing

Similar stones

Rose quartz, milky quartz and quartzite are much harder than halite.

Fluorite has a similar crystal structure, but is harder and its surfaces feel more like glass.

Calcite is harder than halite and has a glassy sheen.

Bronzite ((Mg,Fe)$_2$Si$_2$O$_6$)

A new stone to appear on the market, bronzite is a member of the prolific group of rock-forming minerals called pyroxenes – all complex compositions of metals and silicates. Bronzite derives its name from its warm, metallic, golden-orange lustre. The mosaic-like pattern on the stone's surface hints at its ability to integrate disparate elements or different points of view into one harmonious whole.

Colour:	brown, bronze-brown
Lustre:	vitreous, metallic
Hardness:	5–6
System:	orthorhombic

Identification and care

- Bronzite is always found in massive form with a granular or fibrous appearance. It is often found in association with serpentine.
- It is quite soft and brittle, so may shatter if dropped.
- Unlike metal ores, light plays on its fibres or granules, giving it a mosaic or jigsaw-like pattern.

Magic

- Helps to deflect negative energy
- Promotes creativity and flexible thinking

Tumbled bronzite

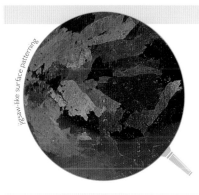

jigsaw-like surface patterning

Healing functions

- Gently warms the body, helping it to relax
- Releases emotional tension
- Aids digestion and absorption of nutrients

Practical ideas

- To dispel nervousness about a new or
  difficult situation: lie down and place bronzite
  at the solar plexus or heart chakra, then
  visualize the forthcoming event. This will
  reduce your anxiety, especially if you place
  a grounding stone (such as smoky quartz)
  beneath your feet and an amethyst at the
  brow (for spiritual calm). Wearing bronzite
  will have the same effect.

Similar stones

Pyrite, or fool's gold,
is similar, but is more
gold/yellow in colour.
Pyrite is also heavier.

Copper has a similar
colour but has no
surface patterning.

77

Staurolite $(Fe_2Al_9Si_4O_{22}(OH)_2)$

This is one of those minerals that can almost look artificial. However, staurolite's inter-penetrating twin prisms are formed naturally. They make cross shapes, and so the stone has become known as "cross-stone" or "fairy-stone". It is often found still embedded in its matrix – usually a soft, creamy yellow rock – making it an impressive display piece.

Staurolite has always been used to bring good fortune and protection. For example, in Switzerland, staurolite crosses were used as amulets at baptisms. In Brittany they were worn as lucky charms said to have fallen from heaven.

Identification and care
- Staurolite is quite a common mineral but not always easily available.
- Its crystals are dull, rough, brown and opaque.

Magic
- Balances opposites
- Establishes poise and stability
- Encourages us to stick to our convictions in the face of opposing points of view

Colour: dark to light brown, yellow

Lustre: dull to vitreous

Hardness: 7–7.5

System: orthorhombic

Similar stones

Basalt crystals are uncommon and are not twinned.

Petrified wood doesn't have clearly defined angles.

Limonite has a characteristic earthy smell.

- Represents the four points of the compass, the elements, the seasons and so, by extension, all the visible world and the sun that illuminates it.

Keywords

Focusing

Quietening

Combining

Healing functions
- Works well at the root chakra and at the base of the spine
- Gently grounding
- Settles the body and mind when energies are overactive and scattered

Practical ideas
- For sleep problems owing to an overactive mind, placing a crystal of staurolite under the pillow can work wonders.

Twinned staurolite crystals   79

Limonite (FeS$_2$ + FeOOH)

Limonite stones are formed in oceanic mud. The surface of the stone is then oxidized through exposure to wind and rain, which produces interesting patterns. Limonite is also known as bog-iron ore since it is found around lakesides.

These strange-looking stones, with their almost organic shapes, can command high prices. The most valuable examples have crystals of golden pyrites showing through the earthy brown layers. These are sold in pairs under the name of Boji™ Stones.

Colour: grey to brown, yellow-brown or black	
Lustre: subvitreous	
Hardness: 5–5.5	
System: orthorhombic	

Identification and care

- Limonite has a pleasant earthy smell.
- It produces a brown streak when scraped.
- It can be found under various names, for example "shaman spheres/balls" and "Moqui marbles".
- The stones are usually roughly spherical or disc-shaped.
- Store limonite in a dry place to prevent excessive oxidization.

A pair of Boji™ Stones

Magic

- Reveals the underlying structures of complex situations
- Helps us to come up with straightforward solutions to problems
- Brings hidden issues to the surface
- Attunes us to nature, nature spirits and spirits of place

Healing functions

- Has a beneficial, clearing effect on the subtle body, creating a sense of well-being and ease
- Boosts physical energy levels

Practical ideas

- For clearing confusion, hold a stone in your hands.
- Place a limonite stone near the sacral chakra (just below the navel) to restore your body's natural centre of gravity.

Keywords

Attuning

Simplifying

De-cluttering

Similar stones

Haematite has a red streak (limonite has a brown streak).

Staurolite forms cross-shaped crystals.

Pyrite (see below) has no earthy matrix.

81

Petrified wood (composition varies)

This is, in fact, not a crystal, but wood that has become fossilized over millions of years to make a powerful combination of tree and stone. Where trees fall into water that has a high mineral content, the wood cells gradually fill with crystalline solutions – most commonly of opal, agate, jasper or chalcedony. Often subtle in colour, with shades of brown, grey and cream, each piece of petrified wood has its own "personality". Like other fossils, petrified wood is a vivid reminder of Earth's long history and of the fact that matter is continually changing from one form to another. Petrified wood lends us the ability of trees to be strong yet flexible.

Identification and care
- Petrified wood can look like wood, but feels like stone.
- Appearance varies depending on mineralization.
- Look for wood grain and cell structure.

Magic
- Increases perseverance and patience

Colour: brown, black

Lustre: vitreous

Hardness: 7

System: triclinic, amorphous, trigonal

Similar stones

Dark-coloured, banded **agate** can look similar, but is less wood-like.

Alabaster has less organic-looking bands.

Jet weighs much less than petrified wood.

Chiastolite can look similar, but it is easily distinguishable by its inner cruciform (cross-like) patterns.

Also: **jasper**, **barytes** (especially oakstone), **basalt**

- Conveys a sense of permanence and of the distant past
- Anchors us to the planet

Healing functions

- Aids memory
- Eases stiffness in joints
- Helps us to understand how events in the past have relevance in the present
- Brings stability in times of stress

Practical ideas

- If a part of your body feels weak, place a piece of petrified wood on it to restore energy.
- To regain emotional strength when you feel "off colour": put tektite at the solar plexus; petrified wood at the sacral chakra; petrified wood either side of the knees; and smoky quartz (point down) below the feet.

Keywords

Memory

Reliability

Permanence

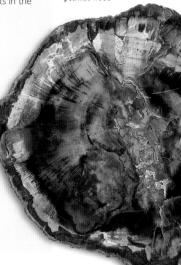

A cross-section of petrified wood

Garnet (composition varies)

Garnet is formed from complex mixtures of minerals resulting in a wide range of colours and compositions. However, it is with the colour red that the stone is usually associated. Indeed, the name "garnet" is thought to come from the Latin word for pomegranate, the seeds of which resemble small pieces of red garnet. Red has always symbolized the blood, and blood has always symbolized vitality in all its forms – such as strength, stimulation, success, fertility and warmth. Garnet seems to embody and promote all these qualities.

Colour: red-brown (pyrope); red-purple or black-brown (almandine); red-orange (spessartine); red or green (grossular); all colours (andradite); dark green (uvarovite)

Lustre: vitreous

Hardness: 7–7.5

System: cubic

Identification and care

- Garnets appear in many colours, each with a different name. The most common is a dark brown-red form called pyrope.
- Crystals are common. They are often roughly spherical with complex multiple facets over all surfaces.
- Garnet crystals are very heavy for their size.

Garnet crystals

- Apparently opaque stones may show deep red through a strong light.

Keywords

Initiating

Stimulating

Fiery

Magic
- Gets projects under way
- Provides fortitude and courage

Healing functions
- The colour will tend to determine the focus of its use for healing. However, all garnets speed up healing processes.
- Red garnets are powerfully energizing – use only for short periods.
- Green garnets are gentler, but still stimulate.
- All garnets bring a fiery, warming energy to fight cold, sluggish or damp disease states.

Similar stones

Ruby, zircon and dark tourmaline are similar to red garnets.

Jade, nephrite and serpentine are similar to green garnets.

Practical ideas
- When using a clear quartz wand to direct energy to a sluggish area of the body, hold a garnet in the palm of the same hand, so that it touches the wand. This will boost the energy flow along the wand and increase the effectiveness of the healing work.

Ruby $(Al_2O_3(+Cr))$

Ruby is a variety of the mineral corundum, coloured by chromium oxides. (The blue variety is sapphire and there are also yellow and colourless corundums). High-quality gem rubies of rich colour and translucence are rare, hence their great value. Much more common are red and purple-blue opaque stones with a heavy metallic sheen. But even low-quality rubies have a mouth-watering beauty that makes them hard to resist. Traditionally the gemstone of the sun, ruby is believed to confer wealth and power, long life and health.

Colour: red	
Lustre: vitreous to adamantine	
Hardness: 9	
System: trigonal	

Identification and care

- Ruby has distinctive barrel-shaped, hexagonal crystals, which narrow toward each end.
- The stone is sometimes cut into a cabochon to show asterism (star-like pattern), caused by inclusions of rutile.
- Hexagonal cross-sections are most commonly used in healing.
- Polished surfaces often show hexagonal growth patterns.

Tumbled ruby

Magic

- Acts like the sun – as a reliable source of life-energy – and brings the wearer benefits associated with the sun (for example, vitality and success)
- Smoothes relationships with other people

Healing functions

- Balances the heart, both subtly and physically
- Helps to engender feelings of confidence, security, enthusiasm and self-esteem
- Warms and steadies

Practical ideas

- To re-establish balance within the Self, use six rubies: place one at the top of the head; one below the feet; one near each knee; and one near each elbow. The effect is amplified if you lie on a yellow cloth.
- To increase your power to achieve a goal, place a ring of 12 quartz crystals evenly around the body, with the points facing outward, then put a ruby over the heart chakra. Lie on a white cloth.

Keywords

Positive thinking

Nourishing

Steadying

Similar stones

Sapphire is similar in shape, but is usually blue or violet in colour.

Zircon, **garnet**, **spinel** and **red tourmaline** can be similar in colour, but rubies (except for those of the highest grade) have a characteristic metallic sheen.

Zincite (ZnO)

The red oxide of zinc, zincite is among a growing number of artificial crystals entering the market. Small hexagonal crystals occur naturally, but are rare. The first large zincite crystals were produced unintentionally as a result of an underground fire in a Polish zinc mine, but most are now created specifically for sale as a by-product of the zinc-smelting process. Unlike some other manufactured crystals, zincite is non-toxic and chemically stable. A happy accident itself, zincite suggests unexpected results and surprises. Its colours are activating and stimulating, and its intense lustre emphasizes its enlivening qualities.

Colour: red, orange, acid green	
Lustre: adamantine	
Hardness: 4–5.5	
System: hexagonal	

Identification and care

- Zincite crystals are often distorted or curved or resemble blunt-ended pencils with many-faceted sides.
- Parts of the crystals may appear melted or fused.
- Crystals are heavy, but quite soft and brittle, so they should be handled carefully and stored separately.

A red zincite crystal

crystals often appear melted

Magic

- Represents novelty, invention, creative experimentation and surprise revelations

Healing functions

- Activates sluggish conditions, such as poor digestion, tired muscles and heavy-headedness
- Helps the body to fight infection

Practical ideas

- Zincite has a fast, effervescent energy that works well with stabilizing stones. Combine it with black or brown stones to activate practical skills; clear or white quartz for cleansing; orange stones for healing; and amethyst for alert mind and inspiration.

Similar stones

Cinnabar, the very soft and poisonous mercury sulphide, has a similar rich red colour, but its crystals are smaller, more vividly coloured and usually on a matrix.

Rutile has a metallic surface lustre and regular striations.

Iron quartz (SiO$_2$)

Also known as ferruginous quartz, iron quartz is found where the quartz has crystallized in the presence of iron oxides. It can look just like dirty, stained, brownish-grey rock crystal, but sometimes the colour becomes a vibrant brick-red or orange.

With its combination of two important minerals, iron quartz is a powerful healing crystal. It brings together the supernatural clarity of rock quartz and the earthy blood-red of iron. Even small, dull iron quartz crystals suggest this intermingling of earthly and spiritual energies.

Colour: red-brown, orange

Lustre: dull to vitreous

Hardness: 7

System: trigonal

Identification and care

- Iron quartz crystals can appear opaque and dull or have areas and veils of reddish colouring.
- Sometimes the external faces of iron quartz crystals may have been partially polished to reveal the colour beneath.
- Iron quartz can also be found under different names, for example as "orange quartz" and "tangerine quartz".

Polished iron quartz

Magic

- Reveals, or makes more tangible, fine or subtle energies
- Encourages practical, effective action
- Accesses the hidden energy of nature

Healing functions

- Grounds us in the here and now
- Optimizes blood circulation
- Helps to restore energy and a sense of well-being

Practical ideas

- If you are feeling unable to focus effectively on what you should be doing, place an iron quartz crystal below your feet and imagine your breath entering and leaving your body through the crystal.
- If your mind is overactive or you are feeling emotionally agitated or upset, place iron quartz at the solar plexus.

Keywords

Gently energizing

Restorative

Condensing

Similar stones

Red jasper displays no visible crystal structure and is rarely translucent.

Carnelian is more richly orange in colour and may have some banding.

Jasper (SiO₂)

Jasper is a type of quartz coloured by a variety of impurities. Haematite makes jaspers red, limonite makes them brown or yellow and chlorite makes them green.

Jasper has a long history, having been used to make amulets and ornaments for many centuries. Its modern name is thought to derive from the ancient Assyrian word for the stone, *ashpu*. Not only was it often red, the revered colour of life, but the endless variety of tone and pattern, and the fact that jasper forms in massive, carvable blocks, made it a valuable commodity.

Colour: red, brown, yellow, green	
Lustre: vitreous to dull	
Hardness: 6–7	
System: trigonal	

Identification and care

- Red jasper is very common. When broken and recrystallized with more quartz solution it resembles a mosaic.
- Many varieties of jasper have been given their own names – for example, there is bloodstone or heliotrope, which is green with red spots, and there is orbicular jasper or "ocean jasper", which features green, blue and yellow concentric patterns.

A block of natural jasper

Magic

- Traditionally was believed to protect from the effects of snake bites and other poisons
- Provides unique solutions to practical problems

Healing functions

- Gently grounds (especially the red and brown varieties)
- Helps to focus on the practicalities of life
- Encourages enthusiasm and drive
- Acts as a steadying anchor when psychic work is undertaken
- Nurtures any damaged areas of the body, aiding recovery and repair

Practical ideas

- To experience the healing qualities of jasper: place one yellow jasper at the centre of the forehead; one yellow jasper between the heart and throat; one red jasper by each ear; one red jasper at the heart; and one green jasper in each hand.

Keywords
Practical
Grounding
Repairing

Similar stones

Iron quartz has fewer colour varieties and is often found as crystals.

Dark carnelian is richer in colour

A tumble of "ocean jasper"   93

Spinel (MgAl$_2$O$_4$)

Spinel rarely forms large crystals, more often resembling the thorns or spines from which it takes its name. From a practical point of view, this can make the stone difficult to use in healing – unless crystals are still within their matrix rock (as is often the case). High-grade spinels have a brilliance approaching diamond, making them popular for jewelry. The stone was traditionally worn as a charm for good luck and to inspire noble thoughts. Like many red stones spinel was also used as a remedy for bleeding.

Colour: red or brown (sometimes blue, green or colourless)

Lustre: vitreous

Hardness: 8–9

System: cubic

Identification and care

- Spinel crystals are usually small octahedral or pyramidal points.
- Look for varieties of spinel under different names: pleonast is black; ruby spinel, blood-red; balas ruby, pink-red; almandine spinel, blue-red; and rubicelle, mauve-red or straw-yellow.

Magic

- Acts as a spur to decisive actions

An octahedral spinel crystal

Healing functions

- Focuses healing energy at the areas of the body that require attention – particularly effective at transferring healing deep into body tissues
- Helps the body to detoxify itself
- Boosts sluggish energy flow
- Loosens stiff muscles

Practical ideas

- Combine spinel and quartz to release blocked energy.
- To alleviate a dull headache, place spinel at the forehead and two or three clear quartz points facing out above the top of the head.
- For digestive cleansing, position spinel on or just below the navel, with two or three quartz points below the feet.
- To increase creativity, place spinel at the sacral chakra, turquoise at the throat, and smoky quartz, point down, below the feet.

Keywords

Initiating

Focusing

Cleansing

Similar stones

Garnet has a multi-faceted, globular crystal shape.

Ruby is more metallic looking and often has larger crystals.

Zircon crystals (see below) are more barrel-shaped.

Carnelian (SiO₂)

Carnelian, or cornelian as it used to be known, is an excellent, all-round healing stone – gently warming and very stabilizing. Few stones are as accessible, as versatile, as "friendly". It works particularly well with the second, sacral chakra. Some may find the energy of carnelian a little too strong, but the addition of other cooling stones will soften its effect.

Carnelian crystallizes from silica-rich solutions coloured by iron oxide impurities. Some of the best sources are India, Brazil, Iran, Saudi Arabia and Uruguay, although a walk along any shoreline is likely to reveal pebbles of this orangey quartz.

Colour: orange, red-orange	
Lustre: vitreous	
Hardness: 7	
System: trigonal	

Identification and care
- Carnelian is densely coloured, translucent to opaque.
- The colour can be enhanced by gentle heat.
- The stone is a form of chalcedony occurring as pebbles or massive lumps.

A carnelian pebble

96

- Beware of imitations: other chalcedony is often dyed to resemble carnelian.

Magic
- A popular stone in carved amulets, rings and seals, worn for courage, confidence and protection from harm

Healing functions
- Releases stress and trauma
- Enhances creativity
- Repairs subtle bodies

Practical ideas
- To speed up healing processes, or to release stress or old trauma, place six carnelians in the following locations: at the throat; above the head; to each side of the body at the level of the sacral chakra; between the legs at mid-calf level; and between the ankles. The effect is amplified if you lie on an orange cloth.

Keywords

Releasing

Repairing

Warming

Similar stones

Sardonyx resembles carnelian, although the orange is often interspersed with layers of white and grey.

Realgar (arsenic sulphide) and **cinnabar** (mercury sulphide) tend to form small, brilliantly coloured crystals on matrix rock. These two stones are both poisonous and should not be used in crystal healing.

Also: dark **citrine** tumbled stones and orange **aventurine**

Sunstone ((Na,Ca)Al$_{1-2}$Si$_{3-2}$O$_8$)

Most people are familiar with moonstone, with its soft, iridescent, pearly lustre. Sunstone belongs to the same mineral family (feldspar), but it contains brightly sparkling inclusions that reflect a warm, golden light.

Found in Norway and the USA, gem-quality sunstone has high proportions of reflective colour, best displayed in cabochons. Lesser grades are largely milky white or brown-orange with occasional flecks of brilliance.

Colour: orange-brown-gold with gold flecks

Lustre: vitreous, adamantine

Hardness: 6–6.5

System: triclinic

Identification and care

- Sunstone is usually found as cabochons or tumbled, and occasionally in faceted form.
- It gives a variable play of light caused by minute haematite or goethite plates, and it always glitters golden-red-brown.
- Synthetic sunstone is made from brown glass and metallic dust.

Magic

- Represents lightness, brightness and optimism

Tumbled sunstone

- Just as moonstone seems to contain the soft lustre of the moon, so sunstone reflects the brighter golden dazzle of the sun

Healing functions
- Warming, gently energizing
- Encourages artistic skills and expression
- Feeds a sense of self-worth and promotes a positive self-image
- Supports a willingness to stand up for who we are

Practical ideas
- To lift sadness and depression: place a sunstone at the solar plexus surrounded by four clear quartz points (points facing out); and a second sunstone at the heart centre with four quartz points (points facing in).
- If you feel feel demoralized by the dullness of a long winter, holding sunstone and looking at its iridescence for a few minutes will help to lift your mood.

Keywords

Warmth

Self-worth

Happiness

Similar stones

Quartzite is white and granular but without iridescence.

Moonstone has a soft blue iridescence.

Spectrolite is white with glassy blue iridescence.

Labradorite (see below) has blue, green, yellow and pink iridescence.

Aragonite (CaCO₃)

Named after the Spanish village, Molina de Aragón, where it was first identified, aragonite has the same chemical composition as calcite, but with a different crystal structure that makes it harder and heavier. This striking crystal conjures up all manner of associations: a futuristic, alien space city; an intrusion from an alternate reality; or the pristine expansion of order into space. Its ability to hold the imaginative mind in a form of creative trance makes aragonite a useful crystal for lateral thinking.

Colour: orange, golden brown, yellow, white

Lustre: vitreous, pearly

Hardness: 3.5–4

System: orthorhombic

Identification and care

- Crystals form long, intergrown six-sided prisms around a common centre. These clusters are called "sputniks".
- Tumbled stones often resemble solidified honey.
- Varieties of aragonite include: iron flower ("flos ferri"), a fine dendritic aggregate; peastone, a mass of pea-sized lumps in a matrix; and sprudelstein, which has different coloured bands like agate.

A "sputnik" cluster of aragonite crystals

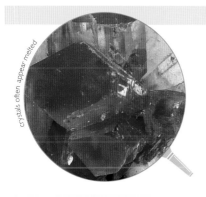

crystals often appear melted

Magic

- Pulls disparate elements together
- Provides a safe base from which to explore new possibilities

Healing functions

- Releases feelings of stress and impatience
- Helps with problem-solving
- Useful for studying, and taking exams or tests

Practical ideas

- For clarity of mind and expression, place aragonite stones at the centre of the forehead, base of the throat and below the navel.
- To steady nerves, place at the solar plexus.

Similar stones

Orange calcite has the same chemical composition as aragonite, but its surface feels different.

Citrine's internal fractures and veils are not as linear as those in aragonite.

Carnelian doesn't have internal fractures.

Copper (Cu)

Copper is a secondary mineral, formed when chemical reactions occur between copper-bearing solutions and other minerals. Natural copper nodules and crystal masses, called native copper, are widespread but not common. Most copper is extracted from ores like chalcopyrite, azurite, cuprite and malachite.

Copper was sacred to Venus, the Roman goddess of love and creativity, who was born on the island – Cyprus – that gives the metal its name.

Colour: orange, orange-red	
Lustre: metallic	
Hardness: 2.5–3	
System: cubic	

Identification and care

- Native copper forms smooth or dendritic nuggets.
- Green copper oxides may be present.
- Very smooth pieces may not have been formed naturally but can still be used for healing.

Magic

- Eases conflicts
- Helps situations to reach a positive outcome
- Promotes emotional harmony

Similar stones

Rutile crystals have a copper colour, but are long and needle-like.

Bronzite crystals are more yellow and have a distinctive mosaic-like surface patterning.

- Alleviates and helps to integrate difficult astrological and cosmic influences

Keywords

Anti-inflammatory

Smoothing

Unifying

Healing functions
- Reduces inflammation
- Releases tension and frustration
- Enhances brain function

Practical ideas
- Wearing copper – for example, in the form of a bracelet – is a well-known remedy for reducing the aches and pains of rheumatism and arthritis.
- Irritation, mood swings, tension and insomnia can be created by the movements of the sun, moon and planets. If you are feeling a little "off" for no apparent reason, carry or wear some copper, or put some under your pillow. If your state improves, then you can be fairly sure that planetary movements were a contributing cause.

A piece of native copper

103

Citrine quartz (SiO$_2$)

Citrine is an uncommon variety of quartz that combines the bright clarity of quartz with the warm tones of sunlight. It occurs naturally where quartz crystallizes with inclusions of iron. The finest specimens come from Brazil, France, Russia and Spain. However, much of the citrine available today is derived from the far more common quartz amethyst, which turns golden yellow when subjected to heat.

Colour: yellow, golden-brown or orange-brown

Lustre: vitreous

Hardness: 7

System: trigonal

Identification and care

- Natural citrine is pale and translucent, sometimes resembling very pale smoky quartz.
- Heat-treated citrine has deep-coloured terminations of yellow, gold or brown, with milky-white bases.
- Ametrine is a variety of quartz that includes yellow citrine areas and violet amethyst areas in the same stone.

Magic

- Sustains all processes in nature
- Incubates energy

Natural citrine quartz from the Congo

Healing functions

- Anchors and activates the lower chakras
- Is restful and warming
- Boosts confidence and creates a sense of personal power
- Provides a "sunny" energy to stimulate the mind in the winter months

Practical ideas

- Use citrine and clear quartz in a layout to enhance communication, adaptability, mental energy and coordination. Lie down and place citrine at the solar plexus, pointing toward the feet. Put clear quartz on the arch of each foot and another above the crown of the head, all pointing upward. Hold citrine in each hand, with the points down the body. The process will be complete when you feel the need to move the stones away, or after ten minutes.
- Citrine and amethyst are complementary in colour and work well together to bring ease, restfulness, balance of mind and optimism.

Keywords

Warming

Comforting

Uplifting

Similar stones

Amber is much softer than citrine and has a slightly sticky feel.

Calcite is much softer than citrine.

Amber (fossil resin, similar to $C_{10}H_{16}O + H_2S$)

Although not actually a stone or crystal, but fossilized resin from coniferous trees, amber has been used for millennia as a precious gem. The main sources of amber are South America, Myanmar and the Baltic coasts of Europe, but it has been found in many places worldwide because it is light enough to float on water. The resin melts easily and is soft enough to be worked with sandpaper.

Colour: gold, yellow, brown, sometimes green

Lustre: resinous

Hardness: 2

System: amorphous

Identification and care
- Amber feels light and is somewhat sticky to the touch.
- Pieces that contain insects may be plastic or reconstituted resin.
- "Russian amber" is made out of small pieces of amber melted together.
- Copal is another type of resin, which is much younger than amber (one or two million years old, rather than over four million).

Magic
- Symbolizes life-energy
- Concentrates the magical energy of the sun

A polished piece of amber

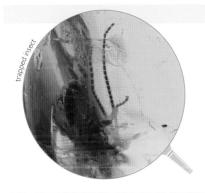

trapped insect

Keywords

Enlivening

Stimulating

Strengthening

Healing functions

- Invigorates the body's systems
- Helps with nervous disorders
- Boosts energy and generates enthusiasm
- Sharpens thinking processes

Practical ideas

- To sharpen thinking, place amber at the brow and hold a piece of jet in the hands.
- Energize the brain and nervous system by lying down, placing jet between the feet and amber above the head, near the base of the spine (between the legs) and in each hand.
- To release nervous tension, run a piece of amber along the spinal column – head to feet – several times.

Similar stones

Orange calcite lacks the sticky feel of amber and is heavier and harder.

Citrine quartz is much heavier and harder than amber and feels cool to the touch.

Pyrite (FeS$_2$)

This stone is popular among many collectors
for its beautiful lustre and fascinating crystals.
Sometimes called iron pyrites or fool's gold,
this common mineral is found in all kinds of
rock and is associated with heated water. The
largest deposits are found in Spain. Its name
derives from the Greek word *pyr*, meaning
"fire" – because it sparks easily when struck
and has been used in fire-making since ancient
times. Its reflective gold colour also links it to
the sun. In healing, pyrite has
a natural affinity to the solar
plexus and abdominal region.

Colour: light brass yellow,
with a dark tarnish

Lustre: metallic

Hardness: 6–6.5

System: cubic

Identification and care

- Pyrite has reflective,
  brassy gold surfaces.
- Crystals often have
  striated faces, and
  the striations always
  run in different
  directions.
- The mineral is found in
  perfect cubes, as well as complex

A cluster of intergrown
pyrite crystals

clusters, massive lumps and radiating striated disks – pyrite "suns".

- Pyrite is soft and brittle, so it is best to keep it away from harder stones.

Magic

- Protects against all forms of darkness
- Can be used as a mirror to reveal the truth

Healing functions

- Cleanses, enlivens and inspires enthusiasm
- Eases anxiety, depression and frustration
- Strengthens logical thinking and creates a grounded sense of reality
- Protects from pollutants and negative energy

Practical ideas

- Large clusters are inexpensive and excellent for brightening up rooms – especially workspaces and eating areas.
- Place pyrite on any area of the body that feels sluggish.
- Pyrite is an ideal support for diets and detoxifying regimes.

Keywords
Activating
Cleansing
Sparking
Clarifying

Similar stones

Gold is generally a darker yellow.

Chalcopyrite is often found with pyrite but tends to be more red in colour.

Pyrrhotite, cubanite and **sperrylite** are all similar in colour and shape but less widely available.

Also: **bronzite, limonite**

Gold (Au)

Gold has epitomized power and wealth for the last 10,000 years. It has long been regarded as one of the most precious and spiritual of substances. Although it is a soft metal, gold is resistant to most corrosive materials and doesn't rust or tarnish. It occurs as tiny grains, or dendritic (branching) shapes. Within lodes, gold is found in thin veins associated with quartz or pyrite. Sometimes these deposits are eroded and the gold washes into streams or rivers, where it settles at the bottom owing to its heaviness – this is a "placer" deposit. Gold also occurs with other metals, such as copper, silver, nickel and lead.

Identification and care

- Small nuggets or flakes of gold in quartz can be used in healing. Gold foil or gold leaf, although worked, can also be used.
- Gold jewelry contains only a small amount of pure gold, so is less useful in healing.
- For easier handling, attach small pieces of gold to thin slices of quartz using silicon glue or beeswax.

Colour: yellow, gold and orange

Lustre: metallic

Hardness: 2.5–3

System: cubic

Similar stones

Pyrite has small flecks that are similar to those of gold, but they are often cubic in shape, whereas gold flecks are more irregular.

Chalcopyrite is a more orange metallic colour.

Electrum, a naturally occurring amalgam of gold and silver, is more silvery yellow.

Magic
- Harnesses the energy of the sun
- Encourages truth and purity
- Boosts wealth, abundance and health

Healing functions
- Balances the brain and nervous systems
- Strengthens the immune system
- Bolsters a sense of self, self-worth and self-confidence
- Promotes creative thinking
- Stabilizes the subtle electrical system of the body

Practical ideas
- Place gold at the heart chakra to infuse the body with life-energy. Surround it with clear quartz points to amplify and speed up the process.
- Place gold at blocked or painful areas to activate healing.
- A gold gem essence is very useful as a healer. Place a few drops on wrist pulses or chakras.

Keywords
Confidence
Creativity
Stability

Gold flakes in quartz (left) and a gold nugget (right)

Rutilated quartz ($SiO_2 + TiO_2$)

Also known as sagenite, angel hair or Venus hair, this beautiful combination of quartz and rutile is found in the Alps in Europe, the Urals in Asia, and in Brazil. Prized as a precious stone, it is cut into facets and cabochons. Rutilated quartz's pattern of criss-crossing crystals suggests rapid communication and exchange of energy. Its clear and orange/gold tones make it excellent for encouraging the body to repair itself.

Colour: clear with fine strands of yellow or orange-brown needle-like inclusions

Lustre: vitreous

Hardness: 7

System: trigonal (quartz) and tetragonal (rutile)

Identification and care

- The body of the quartz is often slightly smoky in colour.
- When the rutile crystals are thin, they appear as brassy or coppery metallic threads, and are usually straight.
- The thicker and more numerous the rutile crystals, the more opaque the stone.
- Larger individual rutile crystals can sometimes be found. They are more silvery in colour, are striated and display a deep translucent red when held up to strong light.

A rutilated quartz point

thread-like rutile crystals

Magic
- Speeds up events and processes
- Removes obstacles

Healing functions
- Helps to knit damaged tissues together
- Relaxes tense muscles
- Lightens heavy moods
- Helps us to see beyond fixed ideas
- Encourages us to make progress, no matter how tough life gets

Practical ideas
- Place rutilated quartz on any parts of the body that have suffered tissue damage, broken bones or muscle strains.

Similar stones

Tourmaline quartz may be similar where the tourmaline is fine, but the inclusions do not have the same brilliant, metallic reflection as rutile fibres.

Goethite is another common inclusion in quartz. It can be found in amethyst as flat, radial, brassy gold crystals.

Tiger's eye is opaque and generally contains more brown than rutilated quartz does.

113

Topaz (Al₂SiO₄(F.OH)₂)

This precious stone has been used in jewelry for centuries, sometimes appearing in amulets of power and influence. Its rarity has meant that it has become associated with glory and success. A long topaz crystal makes a powerful healing wand, able to focus large amounts of energy into the body. Usually thought of as a golden orange or yellow stone, topaz occurs in a wide range of colours, depending on the trace amounts of iron or chromium it contains. Some varieties lose colour in sunlight, while in others the colour gets more intense. Pink topaz is very rare but can be created by heating the yellow variety. Clear topaz can be made blue using heat and radiation.

Colour: orange, yellow, pink, white, blue, white, grey, green, brown, clear

Lustre: vitreous

Hardness: 8

System: orthorhombic

Identification and care

- Topaz can form very large crystals, but these are expensive.
- The highest-quality topaz is rarely tumbled, but white and blue forms occasionally are.
- Crystals can be recognized by their square or rhombohedral cross-section.
- The sides of topaz crystals are striated.

Peach topaz from Pakistan

Imperial topaz
from Brazil

- Topaz is hard, but brittle. Internal fractures can often be seen, especially in heat-treated stones.
- Regularly cleanse crystals worn as healing pendants to preserve their brilliance.

Magic
- Fosters self-assurance and good leadership
- Encourages people to express their desires

Healing functions
- Releases physical tension
- Helps to stabilize the emotions
- Increases motivation and confidence
- Harmonizes all layers of subtle energy

Practical ideas
- Place a crystal at the diaphragm, just below the ribs, to relax the body quickly and to balance upper and lower parts of the body.

Keywords

Assuredness

Confidence

Relaxation

Similar stones

Tourmaline has three-sided crystals, whereas topaz has four-sided crystals.

Brazilianite is similar in colour, but softer.

Aquamarine is similar to blue topaz, but less brilliant.

Also: **heliodor** (yellow beryl), **phenakite**

115

Calcite (CaCO₃)

This is a very common mineral, which forms many types of rock: limestones, marble, travertines and marls, as well as stalactites and stalagmites in cavern complexes. Calcite crystallizes at low temperatures in a wide variety of shapes and colours, and it often forms large twinned crystals. It is also the main constituent in the shells of sea creatures, which means that when they die and sink to the ocean floor, more calcium-rich rocks, such as chalk, are created. Calcite's healing properties are as varied as its appearance: it has a fast, multidirectional energy.

Colour: comes in all colours, as well as in colourless varieties

Lustre: vitreous, mother of pearl, silky

Hardness: 3

System: hexagonal

Identification and care

- Calcite is soft with a silky sheen.
- It has a soapy feel.
- The crystal is found in striking natural shapes, such as "dog-tooth spar" (pyramid-shaped with a sharp point) and "nail-head" calcite (long and thin with a wide, flat head) – both of which are commonly twinned.
- It can easily be confused with many crystals, depending on the colour of the calcite.

A clear calcite crystal from Iceland

- Calcite will split into smooth-surfaced, rhombohedral pieces if pressure is applied – or if it is dropped.

Magic

- Links the individual to the natural cycles of continuous change and transmutation
- Enables effortless flow

Healing functions

- Shifts energy that has become stagnant or very slow-moving
- Soothes and calms agitated emotions
- Quietens the mind
- Creates clarity by removing friction and dissonance

Practical ideas

- Meditate by looking into and through a piece of calcite – particularly spherical or egg-shaped crystals. The stone slows and diffuses light passing through it, so relax your gaze and allow the calcite to calm your mind and still your thoughts.

Keywords

Soothing

Calming

Cooling

Similar stones

Amber has a slightly sticky feel.

Aragonite has a different crystal form.

Clear quartz, blue quartz and **citrine** are much harder than calcite.

Also: **baryte, celestite, danburite, dolomite, magnesite, chabazite**

Mookaite (SiO$_2$)

This stunning quartz mineral has soft, banded areas of red, brown and yellow, resembling layers of coloured sands. Mookaite is a sedimentary rock, which, like flint, has been formed from the opaline, skeletal remains of tiny creatures, deposited in an ancient sea. It is found exclusively in Australia and carries the same range of colours as Aborigine paintings, which derive from natural earth pigments. Mookaite has extremely soothing qualities when it comes to healing. Whenever there is an upsetting or nervous quality to an experience, mookaite will enable a smooth solution to be found. It carries a sense of earthy groundedness and "no nonsense" clarity.

Colour: marbled red and yellow	
Lustre: vitreous	
Hardness: 7	
System: trigonal	

Identification and care

- Mookaite has distinctive yellow and dark red marbling in flowing areas of colour.
- It includes more earthy tones than similar-looking jaspers.

Tumbled mookaite

Magic

- Increases awareness of the life-giving flow of Earth energy
- Encourages harmonious activity

Healing functions

- Stabilizes the lower chakras and stimulates the immune system
- Creates a sense of safety to explore and experience the unknown
- Keeps ideas within the bounds of practical possibilities
- Allows intense spiritual experiences to be integrated into everyday life
- Helps us to achieve our goals in a relaxed and practical way

Practical ideas

- Place at the solar plexus chakra for a sense of relaxed confidence.
- Place at the hips for a feeling of natural balance and grace.
- Place at the brow chakra to improve your understanding of the natural world.

Keywords

Safety

Practical solutions

Smoothing

Similar stones

Yellow and **red jasper** have more distinct edges between bands of colours.

Tiger iron is heavier and has discrete bands of colour, including metallic pyrite or haematite.

119

Heliodor $(Be_3Al_2Si_6O_{18})$

This clean, bright stone is the yellow variety of beryl, coloured by iron. It is often found with quartz and albite. Its name derives from the Greek words *helion* and *doron* and means "gift of the sun". And like a bright early morning, heliodor has a cleansing, opening energy. It is particularly good for clearing away outworn ideas and unhelpful habits, and for resolving confusion. The stone's energy is even lighter and fresher than that of good-quality citrine or lemon quartz, making heliodor well worth using as a healing crystal when you need to make important decisions or changes in your life.

Colour: yellow, lemon-yellow, sometimes with a bluish tinge

Lustre: vitreous

Hardness: 7.5–8

System: hexagonal

Identification and care

- Yellow is one of the rarer colours of beryl.
- Heliodor crystals often have inclusions of other minerals.
- Heliodor can be identified by its hexagonal shape and parallel striations.

Magic

- Symbolizes fresh starts
- Brings intentions to fruition

Heliodor from Russia

Healing functions

- Quietens and restores a nervous system overworked by prolonged stress
- Soothes anxieties and worries
- Encourages optimism
- Brings clarity and furthers understanding
- Enables better use of energy reserves

Practical ideas

- To relax and ease the nervous system, place a smoky quartz at the feet, an amethyst at the solar plexus chakra and a heliodor at the brow chakra.
- To calm the mind, place a smoky quartz at the feet, heliodor at the solar plexus chakra and an amethyst at the brow chakra.
- To increase mental clarity, put a heliodor at the brow chakra (centre of forehead) and a clear quartz on the pulse points of each wrist. When you breathe in imagine the breath entering the body via these stones at the wrist and brow.

Keywords

Refreshing

Restoring

Clearing

Quietening

Similar stones

Hiddenite is more green.

Brazilianite is more brittle and often more acidic yellow in colour.

Amblygonite belongs to a different crystal system and forms short crystals.

Calcite has clear cleavage (fracture) lines.

Chrysoberyl belongs to a different crystal system, and is more gold.

Also: citrine quartz

Sulphur (S)

With its distinctive bright yellow colouring
and interestingly shaped crystals, sulphur makes
a beautiful mineral specimen. Formed from
volcanic gases, it burns easily, creating a bright
blue flame, and this quality has been known
of since ancient times. Sometimes called
"brimstone", sulphur was referred to in the
Bible as an instrument of God's punishment.
Around the 10th century, the Chinese used
it to make the first gunpowder. In healing,
however, its fiery nature can be very useful.
The finest sulphur crystals come from Italy.

Colour: bright yellow, yellow-brown

Lustre: resinous

Hardness: 1.5–2.5

System: orthorhombic

Identification and care
- Sulphur's odour is reminiscent of rotten eggs.
- It is about as soft a substance as it is possible
  to work with in crystal healing. Small crystals
  crumble more easily than large, single ones.
- Keep it apart from other stones.

Magic
- Reveals hidden motivations
- Awakens us to reality
- Helps us to see beyond everyday events

Similar stones

Orpiment is poisonous
to touch and gives off
arsenic vapours, so
should never be used. It
tends to be more orange
and deeper in colour.

122

Healing functions

- Eases digestive disorders and toxin build-up
- Promotes optimism and enthusiasm
- Energizes the rational mind
- Clears away subtle clutter

Practical ideas

- To speed change and clear the mind, place sulphur above the crown chakra, one clear quartz above that and two clear quartz either side of the quartz, pointing outward.
- To aid with detoxification, place sulphur at the second chakra, a clear quartz point just below that, pointing downward, and smoky quartz at the feet.
- To encourage new starts, place sulphur between the feet, a clear quartz just above that, pointing upward, and another clear quartz below the sulphur, pointing downward.

Keywords
Cleansing
Detoxifying
Enlivening

A natural sulphur specimen   123

Peridot ((Mg,Fe)$_2$SiO$_4$)

Peridot is the gem-quality form of the mineral olivine. Visitors to volcanic landscapes often return with rocks encrusted with peridot. It's an important indicator of volcanic activity and can also be found in meteorites (in the form of olivine). Originally called "topazus" by the Romans after an island in the Red Sea where it was found, the Greeks named peridot "chrysolite" (gold stone), a label still sometimes used for the best transparent crystals. It was valued in the ancient world for its power to dispel terror and evil spirits, especially when set in gold.

Colour: green, yellow-green

Lustre: greasy to vitreous

Hardness: 6.5–7

System: orthorhombic

Identification and care

- Large crystals are uncommon – you're more likely to find peridot as small beads or rough chips.
- It has a high lustre that increases when worked.
- Crystals form vertically striated flattened prisms.
- It can be somewhat brittle, so it's best kept separate.

A mass of small peridot crystals on a matrix

124

Magic

- Symbolizes the vibrant energy of the natural world
- Provides protection from harmful forces and influences

Healing functions

- Removes toxins from the body
- Restores emotional balance
- Overrides unwanted patterns of thought
- Enables us to let go of the past
- Helps to strengthen personal identity

Practical ideas

- To stimulate natural healing processes, place four peridot on the body: one at the base of the throat; one by the heart; and one near each kidney (tucked under the back, level with the elbows). Use a grounding stone such as haematite by the feet to facilitate the process.
- When used as a healing pendulum, peridot will release stress and encourage increased levels of perception.

Keywords

Cleansing

Freshening

Invigorating

Similar stones

Forsterite lacks iron in its composition, but otherwise is very similar to peridot.

Fayalite contains more iron and becomes red or brown with a metallic lustre.

Hortonite (magnesium and iron silicate), is a form of olivine that contains more iron than magnesium. It is commonly a brown or black colour.

125

Chrysoprase (SiO₂)

Deriving from the Greek *chrysos* ("golden") and *prason* ("leek"), chrysoprase is a type of chalcedony that often forms in thick veins between a matrix of light, sandy stone. It has a characteristic colour, a dense, translucent apple green, which makes the mineral in its unworked state resemble some kind of confectionery – green almond paste or coconut icing, perhaps. Bohemia was traditionally the richest source, but now Australia produces some of the finest chrysoprase.

Identification and care

- Chrysoprase is found in massive form, veins and concretions.
- As well as its characteristic brilliant apple green, it can be found in a denser, darker shade of green and, in South America, in a pale yellow.
- The crystal is translucent to opaque with a waxy sheen.
- Chrysoprase is cut into cabochons; beware of imitations made of glass or plastic.
- The colour fades in sunlight and heat.

Colour: apple to emerald green

Lustre: resinous to vitreous

Hardness: 7

System: trigonal (microcrystalline)

Similar stones

Emerald of low quality can resemble chrysoprase, but it is usually cut as a faceted stone whereas chrysoprase tends to be cut as a cabochon.

Jade is much harder, with a denser-feeling surface.

Prase often forms visible quartz-type crystals.

Aventurine contains metallic sparkles.

Magic

- Traditionally, a stone of good luck and success
- Enables us to blend with our surroundings – even reputed to confer invisibility
- Conveys the harmonious energy of nature

Keywords
Serenity
Harmony
Calm

Healing functions

- Promotes sound sleep and deep rest
- Supports cleansing and detoxification
- Brings an increased sense of calm and security
- Increases creativity

Practical ideas

- To link with creative forces and find new directions, place six stones in a hexagon: chrysoprase above the head; chrysoprase by the left shoulder; black tourmaline by the left hand; chrysoprase below the feet; chrysoprase by the right hand; black tourmaline by the right shoulder. To amplify, lie upon a black cloth.

A chunk of natural chrysoprase

127

Prase (actinolite quartz) (SiO₂)

Prase is a deep green quartz that owes its colour to fine, needle-like crystals of actinolite – hence, it's now commonly known as actinolite quartz. Prase was a valuable stone for jewelry and carving in the ancient world.

We were lucky enough to meet a lady who showed us the large cluster of prase crystals she had found while walking along a California beach after a storm. It's amazing what stunning crystals you can find just by chance.

Colour: medium green to dark green	
Lustre: vitreous	
Hardness: 7–7.5	
System: trigonal	

Identification and care

- In some stones the actinolite needles can be seen as an inclusion, while in others they are so dense as to give an almost uniform green colour to the quartz.

Magic

- Accesses the healing energies of our surroundings
- Unites the conscious and subconscious
- Blends the internal energies of the heart centre with the external energies of the natural world

Tumbled prase

needle-like actinolite inclusions

Healing functions

- Broadens our perspective
- Breaks down false barriers
- Helps us to feel in tune with our surroundings

Practical ideas

- To reduce feelings of constriction, isolation or alienation, position three prase: one at the brow; one at the heart; and one between the feet. Add four clear quartz points: one above the crown, point up; one below the feet, point down; one in the left hand, point in; and one in the right hand, point out. Visualize your in-breath entering the stones at your feet and moving up through all the stones, then returning down with your out-breath.

Similar stones

Aventurine has sparkles of haematite or mica.

Moss agate is dendritic and organic-looking and has inclusions of green and brown.

Chlorite is another green mineral commonly included in quartz, but it appears as clouds, veils or phantoms, rather than needles.

129

Prehnite $(Ca_2Al_2Si_3O_{10}(OH)_2)$

This attractive milky-green mineral derives its name from the Dutchman, Colonel Hendrik von Prehn, who discovered it in the 18th century. Prehnite is usually seen as radiating masses or botryoidal (like a bunch of grapes) forms, or occasionally as tabular crystals associated with basalts and feldspars. Originally thought to be a variety of emerald or olivine and commonly found in South Africa, it is sometimes known as "Cape emerald" or "Cape chrysolite". Other sources include Australia, the USA, the UK and the Alps.

Prehnite is notable for allowing the gaze to penetrate a little way into its translucent green interior, making it an ideal stone for relaxing the mind and achieving awareness unbound by our usual experience of time and space.

Colour:	yellow-green, clear
Lustre:	vitreous to waxy
Hardness:	6–6.5
System:	orthorhombic

Identification and care

• Prehnite's pale green tones are translucent and have a soft milkiness, similar to moonstone.

A botryoidal mass of prehnite

- Rounded, botryoidal forms are common and can be seen as polygonal "cells" in polished pieces.

Magic
- Unlocks the unconscious mind
- Melts boundaries of time and space, making it easier to access psychic skills such as clairvoyance, remote viewing and channelling

Healing properties
- Eases nervous indigestion
- Soothes and calms the mind and emotions
- Helps to diffuse acute worry

Practical ideas
- Prehnite is an excellent focus for meditation.
- It also makes an ideal surface for scrying, allowing awareness to look into other times and places as well as enabling you to resolve problems. Let your gaze rest in a relaxed way on the crystal. Look "through" rather than at the stone and let thoughts and images come and go as they will.

Keywords

Softening

Alleviating

Reassuring

Similar stones

Peridot is more lustrous and clear, without prehnite's milky softness.

Wavellite forms flattened, silky, radiating fibres or spherical aggregates.

Hemimorphite (calamine) is found in radiating aggregates, and gives a silky play of light.

Stilbite (bundle zoisite) forms bow-tie shaped bundles of white, yellow or brown crystals. It is soft and brittle with a pearly lustre.

Aventurine (SiO$_2$)

A common variety of massive quartz, aventurine has a long history of use for carving and healing. For example, it was used in Tibet to increase perception and improve eyesight. It can be easily distinguished from other members of the quartz family by its fine, sparkling inclusions. It is this quality that gives the stone its name – it was considered to resemble a sparkly green ornamental glass of the same name, invented in Italy during the 18th century, in which metallic flakes were added to the molten mixture.

Colour: green with metallic flecks (also red-brown, blue)

Lustre: vitreous

Hardness: 7

System: trigonal, microcrystalline

Identification and care

- Available as tumbled stones or rough lumps, aventurine is easily identifiable by its sparkly appearance.
- Deep-coloured, sparkling stones tend to be used for carving, while many tumbled stones are pale with little play of light.

A chunk of natural aventurine

- Aventurine is found in three main colours: green, coloured by green mica; red-brown, coloured by pyrite or goethite; and blue, coloured by rutile or crocidolite.

Healing functions
- Calms and stabilizes the emotions
- Encourages a positive outlook
- Increases tranquillity (particularly blue aventurine)
- Green aventurine is one of the best balances for the heart chakra, and promotes spiritual growth and gratitude for what you have.
- Red-brown aventurine brings happiness and relaxation.

Practical ideas
- To stabilize the emotions and open the heart: place an aventurine in the centre of the chest surrounded by four clear quartz (points facing outward). If your emotions are disturbed, place another aventurine at the base of the throat and a smoky quartz or another grounding stone below the feet.

Keywords

Balance

Tranquillity

Stability

Similar stones

Jade has a less shiny lustre than green aventurine and does not sparkle.

Emerald can, like green aventurine, contain shiny mica, but has noticeable cleavage planes.

Amazonite has tones of green, but does not glisten.

Blue quartz is similar to blue aventurine, but with no sparkle.

133

Emerald (Be$_3$Al$_2$Si$_6$O$_{18}$)

Emerald is the precious green variety of the mineral beryl. It forms in granite and pegmatite rocks, taking its colour from additional chromium and vanadium atoms. The range of colours goes from a plastic-looking milky grass-green to the richest of transparent greens like deep woods in springtime. It is no wonder that emerald has always had a magical association with the plant kingdom and with fertility and abundance. Mined in Egypt 4,000 years ago, today emerald is found mainly in Brazil, Colombia, Russia, Pakistan and southern Africa.

Colour:	bright green
Lustre:	vitreous
Hardness:	8
System:	hexagonal

Identification and care

- High-quality, transparent stones are rare and go to the gem trade, but opaque emerald with cloudy inclusions is easy to find.
- Emerald crystals are long prisms, clearly hexagonal in shape with faceted, domed terminations.

A cluster of emerald crystals

- Lower grades may be called "green beryl".
- Inclusions can identify origin: for example, stones containing calcite or pyrite come from Brazil; those with mica, from Russia or Africa.
- Cut emerald is widely imitated – for example, "Indian emerald" is actually dyed quartz.

Magic
- Reveals the truth, so encourages honesty
- Bolsters love, friendship and attachment
- Associated with prophecy and foresight

Healing functions
- Speeds cleansing and purifying processes
- Assuages hidden fears and anxieties
- Effective as a focus for meditation

Practical ideas
- Place emerald at the heart chakra to bring balance and calm.
- To experience the full benefits of the stone, position six emeralds: on the brow; on the upper chest; on the heart chakra; on each side of the waist; and between the feet.

Keywords
Calming
Cleansing
Harmonizing

Similar stones

Aquamarine is a more blue-green variety of beryl.

Apatite has a "crackled" appearance and is softer.

Tourmaline forms in triangular, rather than hexagonal, prisms.

135

Bloodstone (SiO₂)

The traditional names of the quartz family can be a little misleading, making the varieties seem more different from each other than they actually are. Bloodstone is essentially green jasper that has the addition of red, brown or yellow markings created by iron oxides. However, this tiny variation brings quite a range of healing qualities, as well as making the best specimens – with distinct blood-red markings against a deep green base – a sought-after carving material.

Colour: green with red, brown, or yellow markings

Lustre: vitreous

Hardness: 7

System: trigonal

Identification and care

- Good-quality bloodstone is used in the jewelry trade, so lesser grades with smaller areas of colour are easier to find.

Magic

- Offers protection from loss of life-energy
- Helps to increase our worldly influence and prestige

A palmstone of bloodstone

- Is traditionally linked with healing wounds and stopping blood-loss – a warrior's stone, under the influence of the planet Mars

Healing functions

- Stimulates the physical systems of the body, particularly blood circulation
- Energizes and balances heart and root chakras
- Provides motivation and encouragement
- Encourages practical use of clairvoyant skills
- Brings strength and good health

Practical ideas

- A large bloodstone is useful for meditaton and other spiritual practices. It encourages the flow of energy and strengthens the core consciousness of the individual.
- For increased strength, determination and vitality, use five bloodstones: one close to the base of the spine; one between the heart and throat chakras (the thymus area); one at the forehead; and one on each side of the head, close to the ears.

Keywords

Courage

Strength

Support

Stimulation

Similar stones

Green jasper is identical to bloodstone in composition but has no red flecks.

Chlorite (including seraphinite)

$(Ca_2(Mg,Fe^{2+})_5Si_8O_{22}(OH)_2)$

Chlorite often forms green, cloudy inclusions within quartz crystals, but its most impressive form is known as "seraphina" or "seraphinite". This relatively recent addition to the crystals market, originating in Siberia, forms very soft, radial clusters of dusty green fibres. When carefully polished it resembles crushed velvet or shot silk in its magical complexity of shades and play of light. Initially rare and extraordinarily expensive, seraphinite is now easier to acquire – but, because of its softness, beware of it as jewelry. Many newly discovered minerals are extremely beautiful, but no match for the durability of traditional gemstones.

Colour: pale and dark green to black	
Lustre: pearly to vitreous	
Hardness: 2	
System: monoclinic	

Identification and care

- Chlorites are very soft – keep stones separate from each other.

Magic

- Enables contact with the forces of the natural world, nature spirits and non-physical beings from this planet and beyond

Similar stones

Malachite rarely offers a play of light on its nodular surface.

Astrophyllite has shiny, radiating and overlaid flat splinters of dark orange, green, gold or brown.

- Works on the fine levels of the heart chakra, refining our relationship with the spiritual forces of the universe

Keywords

Opening

Sensitizing

Revealing

Healing functions
- Gently calms overactivity in any body area
- Alleviates feelings of isolation and abandonment
- Clarifies personal goals

Practical ideas
- Hold a piece of chlorite to attune to nature or open up to meditative states.
- Place seraphinite on the heart to access subtle energies or realms, with a grounding stone on the brow to clarify and stabilize the experience. To link to Earth energies, use black tourmaline as a grounding stone; for cosmic energies, use moldavite; and to access angelic realms, use celestite.

Tumbled pieces of seraphinite

139

Dioptase ($Cu_6Si_6O_{18}.6H_2O$)

This striking crystal is a copper silicate found in veins of copper oxidized by air or water. Dioptase is relatively soft for a gemstone but it has great "fire", that is, it has a remarkable dispersal of light rays and play of colours, so that transparent stones are highly regarded – and rare. The intense green colour stimulates all levels of body, mind and spirit, particularly enhancing subtle perceptions and psychic abilities. Dioptase is also an appropriate stone to use when initiating new projects and stages of growth. You may find that its rapid cleansing action is best administered in small doses – hold or wear the stone for only a few minutes at a time, especially if close to the heart chakra.

Colour: viridian (deep blue-green)

Lustre: vitreous

Hardness: 5

System: trigonal

Identification and care

- Dioptase forms rather small crystals of an intense green colour. They are usually squarish, six-sided prisms.
- Individual crystals are less common than small clusters still on their matrix.

Clusters of natural dioptase

striking play of light

Keywords

Stimulating

Nourishing

Expanding

Magic
- Stimulates change and growth

Healing functions
- Nourishes the heart chakra
- Clears away outworn behaviour patterns
- Brings clear vision and penetrating insight

Practical ideas
- To nourish the heart chakra, place dioptase for a few minutes above the heart chakra.
- Keep a piece of dioptase beside your bed to encourage change.
- To help you to achieve a goal, put dioptase next to a drawing or photograph that represents what you are aiming for.

Similar stones

Atacamite is a similar copper mineral with long, needle-like crystals of dark green, often radial, clusters.

Euchroite has short prisms or thick tabular crystals of emerald green. Unlike dioptase, it does not have "fire" or defined edges. It is also poisonous, so wash your hands thoroughly after handling.

141

Diopside (CaMg(SiO$_3$)$_2$)

Diopside is a member of the pyroxene group of minerals, common in calcium-rich, metamorphic rocks. Crystals are fairly rare – they are usually found as granular aggregates. The finest and most precious form of diopside is the "black moonstone", which when cut as a cabochon shows a star of light (asterism) or a cat's eye (chatoyancy). Traditionally, black moonstone offers protection from harmful influences and the "evil eye". Diopsides of all types stimulate new cycles of growth and healing from deep within the body. They can also begin to reverse negative emotional and mental states.

Colour: black, dark green or green (occasionally colourless, yellow or blue)

Lustre: vitreous to dull

Hardness: 6

System: monoclinic

Identification and care

● Varieties of the crystal include the deep green chrome diopside; black diopside; and the rare, violet-blue violane diopside.

A chrome diopside crystal

- Dark green or black varieties of the stone are often tumbled.

Magic
- Reveals hidden things
- Represents the power of the dark moon, associated with magic and the unconscious

Healing functions
- Releases deeply entrenched stress
- Restores energy after chronic illness
- Increases a sense of self-worth
- Helps us to see the best in things

Practical ideas
- To aid recuperation, put diopside on the affected part of the body, clear quartz above the head, and smoky quartz below the feet.
- After stress or trauma: place clear quartz above the head and diopside (one showing asterism is best) one metre below the feet.
- To boost self-esteem: position diopside at the solar plexus; rose quartz at the heart; and smoky quartz at the feet.

Keywords

Revelation

Growth

Renewal

Similar stones

Tourmaline – tumbled tourmaline often retains uneven surfaces.

Jade is translucent or opaque when cut.

Black obsidian is usually translucent.

Dioptase gives a light blue-green streak, whereas that of diopside is white.

143

Epidote $(Ca_2(Al,F^{3+})_3Si_3O_{12}(OH))$

Almost identical in structure and appearance to tourmaline, epidote was not identified as a different mineral until the early part of the 20th century. One way in which it certainly does differ from tourmaline is in its comparatively small range of colours, the most popular of which being a yellow-green reminiscent of pistachio nuts (and so called pistacite). Good sources of epidote are Austria, Sri Lanka, Alaska and Peru. In healing, epidote is valued for its sturdy, buttressing energy.

Colour: pistachio green, green, brown, black

Lustre: vitreous to dull

Hardness: 6–7

System: monoclinic

Identification and care
- Dark green epidote is cut and polished, other types are tumbled.
- Withamite is a bright red to pale yellow variety.
- Epidote is rarely transparent, but some gem-quality green stones are used in jewelry.

Magic
- Establishes, develops, supports
- Suggests practical ways to succeed

A natural epidote crystal

pronounced striations

Healing functions

- Provides a grounding, practical energy, which helps to strengthen the structures of the body, including bones, cartilage, tendons and muscles
- Steadies the emotions
- Helps us to achieve goals
- Enables us to accumulate strength on all levels

Practical ideas

- To achieve a goal, place epidote at the centre of the forehead and in each hand.
- To make practical progress, put epidote between the feet and citrine quartz at the solar plexus.

Similar stones

Tourmaline is very similar in form and so difficult to distinguish from epidote, except that it can be found in a wider range of colours.

Vesuvianite has less "clean"-looking edges to its crystal forms.

Actinolite has a silky lustre and thin needle-like crystals.

Augite has squarish black prisms.

145

Jade [Jadeite: $NaAl(Si_2O_6)$; Nephrite: $Ca_2(Mg,Fe)_5Si_8O_{22}(OH)_2$]

Jade is found in two forms: nephrite (an actinolite), and jadeite (a pyroxene). Nephrite, whose name derives from its supposed ability to heal the kidneys, is not the hardest mineral, but it is the toughest, i.e. the most difficult to break. Jadeite is more granular than nephrite – when polished it looks slightly dimpled.

The reverence for jade is so widespread and its attributes among different cultures is so consistent that a very ancient origin for its use is likely. It is universally considered to be an important healing stone with particular relevance in all matters concerning ancestral spirits and the protection of the dead.

Jade can be found in a variety of colours. For example, a source of light blue jade has been found in California. It also changes colour, often to shades of brown, when it is buried with the dead.

Identification and care

- Imitations of jade, using minerals such as bowenite, grossular garnet, and prehnite, are common.

Colour: green, white, grey, blue-ish

Lustre: greasy

Hardness: 6.5–7

System: monoclinic, microcrystalline

Similar stones

Bowenite is a form of serpentine, which – unlike jade – is soft enough to be scratched with a knife.

Grossular garnet is often labelled "African jade" or "Transvaal jade".

Also: **chrysoprase, prehnite, aventurine**

Magic

- Associated with longevity
- Helps to connect living people with their dead ancestors and to protect the spirits of the dead

Healing functions

- Balances the heart chakra, which helps to improve our relationships with other people
- Increases a sense of belonging
- Enhances the ability to act appropriately and effectively
- Relaxes us at a deep level, accelerating healing processes

Practical ideas

- To re-energize yourself when you feel disconnected from your past or where you live: place jade at the base of the spine, sacral chakra or throat.

Tumbled jade

147

Sphene (CaTiSiO$_5$)

The transparent, gem-quality variety of the mineral titanite, sphene frequently occurs in metamorphic rocks, particularly in cracks within crystalline schists with quartz, chlorites and feldspars. A dirty, translucent olive-green in colour, sphene forms strange-looking, flattened wedge-shaped crystals – indeed, its name means "wedge" in ancient Greek. Like a real wedge, sphene can be used to open the door to new levels of experience or to gain access to deeply entrenched conditions, making it an important healing stone.

Colour: yellow-green, green-brown, brown-red

Lustre: adamantine

Hardness: 5–5.5

System: monoclinic

Identification and care

- The colour is rarely "clean" – usually a brown-green or with reddish tints – but is still very reflective.
- The flat, tabular crystals of sphene are often slightly striated.
- Sphene is sometimes used as an imitation for more valuable stones such as topaz, heliodor and olivine.

A sphene crystal

148

Magic

- Clarifies new directions
- Offers solutions to problems
- Encourages lateral thinking

Healing functions

- Gently increases levels of physical energy
- Helps us to understand our emotions
- Makes us open to new ideas and concepts
- Aligns the subtle bodies

Practical ideas

- For clarity of mind, place sphene around the head or at the brow chakra.
- To increase energy and focus, put sphene just above the upper lip or just below the lower lip.
- Combine with rose quartz at the heart chakra to clear emotional stress.
- Combine with citrine quartz at the solar plexus to increase your powers of discrimination.
- Combine with garnet at the root chakra for motivation and drive.

Keywords
Clarifying
Aligning
Exploring

Similar stones

Olivine is a clearer green and its crystals are less flat.

Peridot (see below) is a clear, transparent, apple green.

Serpentine $(Mg_6(OH)_8Si_4O_{10})$

The name serpentine derives from the ancient belief that the stone was a remedy against snake bites. This idea may have stemmed from serpentine's sinuous patterning, as well as its colouration reminiscent of snakeskin. It is metamorphic, forming from olivine-rich rocks, and is principally found in the UK (in particular, Cornwall), Italy, South Africa, Russia and the USA. The fibrous form of serpentine is asbestos (also known as chrysotile), which should not be used, because its dust can cause serious lung diseases, such as emphysema. However, the massive form of serpentine is completely safe.

Colour: shades of green, yellow, red and brown

Lustre: waxy to vitreous

Hardness: 3–4

System: monoclinic, microcrystalline

Identification and care

- Serpentine is easy to identify by its slippery, soapy feel.
- Bowenite is a pale variety, often called "new jade".
- It is soft enough to be scratched by other stones, so should be stored carefully.

A slice of serpentine

150

Magic

- Has the power to avert harm
- Allays fears
- Develops psychic skills

Healing functions

- Enables absorption and use of *prana* (life-energy) throughout the body
- Calms and stabilizes the mind
- Heightens perception
- Good for meditation and the harmonious integration of new states of awareness

Practical ideas

- Consider using serpentine when anxieties and fears prevent proper relaxation and rest. Nightmares, insomnia, paranoia and fear of the unknown can be eased by wearing or holding this stone.
- To calm a restless mind whirring with thoughts: place serpentine at the throat and heart; carnelian at the sacral chakra; smoky quartz below the feet; and an amethyst above the head.

Keywords

Comforting

Flowing

Enlightening

Similar stones

Epidote feels much harder and has less colour when in its tumbled form.

Jade has a greater hardness and much tighter "grain".

151

Malachite $(Cu_2CO_3(OH)_2)$

A secondary mineral of copper deposits, malachite is often found alongside azurite. Single crystals are rare; it more commonly forms nodules or radiating masses. Malachite contains about 60 per cent copper, so is a useful ore. Large deposits are in Russia, Africa and Australia. The name derives from the Greek for the mallow plant "maloche", which has leaves of the same soft, deep green. Because of its striking concentric eye patterns, malachite has long been used as a magical protection against the "evil eye".

Colour: green with dark green or black concentric bands	
Lustre: silky or dull	
Hardness: 3.5–4	
System: monoclinic	

Identification and care

- Polished pieces, slices, nodules and granular masses are common, and often found with blue azurite.
- It has a light green streak, which distinguishes it from similar stones such as rosasite and pseudomalachite.

Tumbled malachite

- Malachite is very soft and absorbent, so should be kept separate from harder stones and corrosive materials such as salt and oils.

Magic
- Associated with growth and abundance

Healing functions
- Relieves pain and aids recovery after exhaustion
- Locates and corrects emotional imbalances
- Promotes the development of new ideas
- Encourages and strengthens in difficult circumstances
- Absorbs environmental pollutants

Practical ideas
- To conteract long-term exposure to electromagnetic radiation, regularly take two or three drops of the gem essence or hold a piece of malachite in each hand.
- Hold a piece of malachite over painful body areas. Several pieces may need to be used to absorb the imbalances. Cleanse thoroughly.

Keywords

Balancing

Soothing

Detoxifying

Absorbent

Similar stones

Serpentine has less regular patterning, and is more yellow-green in colour.

Chrysocolla is more light blue-turquoise, and has no regular banding.

Rosasite forms botryoidal crusts of green or blue. Its streak is a pale blue.

Pseudomalachite has a blue-green streak.

Moldavite (complex glass/silicate, SiO₂)

This amorphous tektite (see pages 68–69) is found only in one area of the Czech Republic. Moldavite was created 15 million years ago but its origin is a mystery. It may have been formed from the impact of a meteorite melting surrounding rock – it certainly doesn't contain the usual inclusions of volcanic rock. It may even have arrived with a comet or meteorite and so be extraterrestrial. Moldavite has always been accorded special significance. Medieval references to the Holy Grail being a stone that fell from heaven has led some to identify moldavite as that much sought-after relic.

Colour: translucent green, brown-green	
Lustre: vitreous	
Hardness: 5	
System: amorphous	

Identification and care

- Moldavite is the only tektite with a green transparency; the others are black or brown.
- Larger pieces are rare. Most moldavites have pitted, rippled or cratered surfaces and form extruded, teardrop or pebble shapes.

Magic

- Blends cosmic and earthly energies, expanding awareness of the universe

A pendant made from natural moldavite

a typically cratered surface

Keywords

Transforming

Expanding

Amplifying

- Represents transformation, evolutionary processes and spiritual growth

Healing functions
- Amplifies properties of other stones
- Strengthens the immune system
- Heightens our appreciation of creation

Practical ideas
- To encourage innovation or explore your true potential: lie upon a green cloth with celestite above the head, moldavite beside each shoulder, celestite beside each hand, and moldavite between the feet. Make sure that you have grounding stones, such as haematite or tourmaline for use after the session.

Similar stones

Other **tektites** are similar but lack the rich green colour.

Verdelite (green tourmaline) resembles cut moldavite, but is softer and has no dichroism.

Peridot is usually more yellow-green.

155

Moss agate (SiO₂)

Moss agate is a translucent chalcedony quartz with inclusions of green and brown manganese oxides, hornblende and iron. The Red Sea coast of Yemen has an ancient source where the stone was called "mocha stone" after the nearby trading town of al-Mukna.

Holding a slice of moss agate is like having the concentrated energy of nature in your hand: it resembles a section through a fern-fronded forest, a bird's eye view of a river estuary, or a vision of Earth from outer space.

Colour:	clear with green and brown inclusion
Lustre:	vitreous
Hardness:	6.5
System:	trigonal, microcrystalline

Identification and care

- Fern, moss and leaf patterns are clearly discernible in moss agate, although you may need to hold denser stones up to the light.

Magic

- Attunes us to the natural world
- Represents tenacity and steady progress

Tumbled moss agate

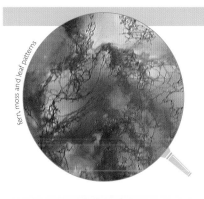

fern, moss and leaf patterns

Keywords

Feeling of space

Optimism

Opening

Healing functions

- Helps to relax and open up congested or constricted areas – for example, in the circulation of blood or lymph
- Releases pent-up emotions
- Increases optimism
- Boosts our sense of our own potential

Practical ideas

- To relieve feelings of tightness, congestion and emotional tension, particularly those concerning personal relationships, place five moss agates on the body: one at the upper chest (thymus); one at the heart chakra; one at the base of the stomach; and one either side of the heart chakra stone.

Similar stones

Tree agate is a milky-white quartz and not very translucent.

Green jasper is completely opaque.

Dendritic sandstone/limestone has the same inclusions as moss agate, but within granular, opaque rock, not quartz.

157

Turquoise $(CuAl_6(PO_4)_4 (OH)_8.4-5H_2O)$

Early Europeans gave this stone the name
pierre turquoise ("Turkish stone") in the belief
that it came from Asia Minor. However, this
was simply where Europeans first traded
stones originating from desert regions such
as Iran, Sinai and Tibet. Turquoise is formed by
water acting on aluminium and copper, which
gives the stone its blue colouration. Shades of
green come from iron impurities.

The stone has always been popular as a
protective amulet. In Renaissance Europe no
gentleman would be seen without a turquoise
set in a ring to ward off riding or duelling
accidents. In many cultures the sky blue
turquoise is combined with an earthy
red or orange stone, such as coral or
carnelian, to harness the combined
protection of heaven and Earth.

Colour: turquoise, turquoise-green, light blue	
Lustre: waxy to vitreous	
Hardness: 5–6	
System: triclinic, microcrystalline	

Identification and care

- Crystals of turquoise are rare
  and usually small. It mainly
  forms in masses or nodules.
- Sunlight will cause turquoise to

Tumbled turquoise

fade. Most samples are impregnated with wax or resin to stabilize the colour.

- Turquoise is porous, and can be discoloured by oil and grease.

Magic

- Enhances and strengthens the aura
- Connects to the spirit worlds

Healing functions

- Strengthens all the organs
- Balances all the subtle systems, particularly the heart, thymus and throat chakras
- Neutralizes environmental negativity
- Cools emotions
- Calms overactive thoughts
- Enhances intuition and psychic skills

Practical ideas

- To release hurts and bring about change in the face of obstacles: place one turquoise above the head; one below the feet; one by each hand; and one by each shoulder. Lie on a yellow cloth to amplify the effects.

Keywords

Strengthens

Protects

Enhances

Similar stones

Howlite, a white/grey amorphous stone, is often dyed to mimic turquoise.

Amazonite is less blue, with bands of colour.

Also: **larimar**, **chrysocolla**

159

Aquamarine ($Be_3Al_2Si_6O_{18}$)

One of the most popular light blue gems,
aquamarine is a variety of beryl. Impurities of
iron determine its colouring, which can range
from almost colourless to deep sea blue-green.
High-grade aquamarine is often imitated with
blue topaz, coloured glass or heat-treated
low-grade stones. Aquamarine is found in
pegmatite, a type of slow-forming igneous rock,
and good sources include Brazil and Pakistan.

Colour: almost
colourless, pale
blue or turquoise

Lustre: vitreous

Hardness: 7.5–8

System: hexagonal

Identification and care

- Aquamarine forms hexagonal pencil-
  shaped crystals.
- Jewelers value specific shades; green-
  tinged stones tend to be less expensive.
- Tumbled stones are pure, milky blue
  to a pearly, turquoise green.
- Parallel striations catch the light
  and depth of colour.
- Crystals show dichroism
  – seen from some angles
  they look colourless.
- Terminations are either flat
  or are faceted domes.

Aquamarine
crystal

Magic

- Protects from harm
- Promotes safe travel, especially at sea

Healing functions

- Boosts the immune system, balances the thymus and throat chakras
- Clears stagnant emotions
- Encourages optimism
- Enables creative expression of ideas
- Encourages unique skills
- Helps you to find inspiration and enables the imagination to flow, usually leading to practical outcomes

Practical ideas

- When your energy is low and infections are frequent: wear an aquamarine pendant, ideally midway between the throat and heart chakras.
- Combine in a layout with rose quartz to calm and soothe on all levels.

Keywords
Cooling
Clarifying
Soothing
Lightening

Similar stones

Blue topaz is a heavier, more metallic blue.

Siberian quartz has no internal features and no dichroism.

Aqua aura is an enhanced clear quartz that is more iridescent.

Tumbled aquamarine

Amazonite (KAlSi$_3$O$_8$)

This popular gemstone is a variety of the common mineral feldspar. It creates attractive, and somehow deeply satisfying, chunky, prismatic crystals of a blue-green colour. The stone is named after the Amazon River, although no deposits are known in that area. It is found mainly in Canada, Madagascar, Mexico, Russia and parts of Africa. Amazonite is relatively soft and easy to work, and it was popular in ancient Egypt. The green colours of amazonite have traditionally been linked to fertility, protection and the spirits of the dead.

Colour: pale turquoise to deep mid-green, with lighter striations	
Lustre: vitreous	
Hardness: 6–6.5	
System: triclinic	

Identification and care

- Its depth of colour varies greatly, but, when polished, amazonite can usually be identified by its streaky, narrow grain with interspersed spots of darker or denser colouring.
- The surface may appear iridescent.
- Fine, clearly defined crystals are particularly attractive.
- When used as a gemstone, amazonite is often cut into cabochons.

Tumbled amazonite

Magic

- Helps you to access distant memories, even of past lives
- Taps into ancestral energies

Healing functions

- Eases problems with the ears, nose, throat and nervous system
- Inspires creativity and personal expression
- Releases blocked emotions
- Supports communication
- Enhances psychic faculties

Practical ideas

- To strengthen self-identity and personal power: place one amazonite at the solar plexus chakra, one at the base of the spine and one beneath each foot.
- To access deep memories: place one amazonite stone at the heart chakra and one at the base of the skull.

Keywords

Stimulation

Communication

Memory

Similar stones

Turquoise has a more even colour, is not banded, and is usually more blue.

Aventurine is not streaky.

Emerald is more translucent and vitreous.

Also: **chrysocolla**

163

Chrysocolla $(Cu_2H_2Si_2O_5(OH)_4)$

Often resembling ethereal landscapes or the surface of distant planets, chrysocolla is a stone that suggests many possibilities. It actually forms where copper is exposed to air and water. The mineral is rarely pure and its crystal structure varies depending on the mix of elements. It is also found mixed with other copper minerals, such as azurite, malachite, cuprite and turquoise.

Colour: green and turquoise to light mid-blue

Lustre: vitreous to waxy

Hardness: 2–4

System: monoclinic or orthorhombic, microcrystalline

Identification and care

- Pure chrysocolla is very soft, unless mixed (naturally or artificially) with minerals like quartz, chalcedony or opal (gem silica).
- High-quality chrysocolla often has distinct areas of blue and green that make it look like the Earth from space.
- Eilat stone from Israel includes equal amounts of turquoise and malachite. It has a more consistent colouring than most other types of chrysocolla.

Magic

164

- Brings change, success and peace

Tumbled chrysocolla

the surface is like Earth viewed from space

Keywords

Absorbing

Communicating

Relaxing

Healing functions

- Good for throat and chest problems
- Relaxes the body and emotions
- Enhances the flow of creativity
- Aids communication on many levels

Practical ideas

- To relax breathing and increase expressive ability: place chrysocolla at the base of the throat and on each side of the ribcage.
- To access subconscious information: place one chrysocolla at the throat chakra and one at the base of the skull.
- To balance energy flow in the spine: place chrysocolla at the base of the spine and at the base of the skull.

Similar stones

Larimar contains white veining and may be harder.

Turquoise has a more even colour and is harder.

Amazonite is harder.

Blue lace agate is harder, more blue and does not contain green patches.

Larimar (Ca$_2$NaSi$_3$O$_8$(OH))

This striking form of the mineral pectolite is a subtle turquoise blue with soft, white veining, reminiscent of tropical seas. It is rare, being found in only one location, in the mountains of the Dominican Republic. It occurs in radiating aggregates of fine needles and as compact masses filling cracks and cavities in basalt.

A spiritual master once claimed that the islands of the Caribbean were part of the lost continent of Atlantis. He said that the area would produce a blue stone with the power to heal. Some believe this referred to larimar, and it is sometimes called the "Atlantis stone".

Colour: light turquoise, light blue, white

Lustre: silky to vitreous

Hardness: 4–5.5

System: triclinic

Identification and care

- Larimar is rare, so quite expensive.
- Larimar is also sold as "dolphin stone".
- It is a soft mineral, keep it away from harder stones and sharp edges.

A mass of larimar in basalt

Healing functions

- Aids the flow of *prana* (life-energy) in the upper body and releases any tightness there
- Balances extremes of emotion
- Encourages tolerance
- Gives an appreciation of personal space
- Relaxes the conscious awareness
- Broadens experience of time and space
- Helps you to perceive your place in the overall scheme of things

Practical ideas

- To relax and broaden your awareness: place larimar at the brow; two clear quartz points around the head (pointing outward); and a grounding stone between the feet.
- To bring peace and contentment: place a blue stone at the throat chakra; larimar at the heart chakra, with two clear quartz points either side (pointing outward); a citrine quartz at the solar plexus chakra; and a smoky quartz between the legs.

Keywords

| Far horizons |
| Equanimity |
| Acceptance |

Similar stones

Hemimorphite (calamine) appears to have a "silky" sheen.

Also: **turquoise**, particularly pale varieties

167

Smithsonite ($ZnCO_3$)

Many collectors have been attracted to smithsonite by its unusual silky, pearlescent sheen. But it is intriguing also for its typical form, which is botryoidal – resembling a luscious bunch of grapes. In ancient times smithsonite was a source of zinc, which was mixed with copper to make brass. This ability to blend easily reflects its powers to enhance creativity and aid smooth transitions. The stone is named for the British scientist James Smithson, who founded the Smithsonian Institution.

Colour: pink, blue, green, blue-green, grey

Lustre: vitreous to pearly

Hardness: 4.5

System: trigonal

Identification and care

- Large crystals are rare. Smithsonite usually forms rounded (botryoidal) clusters and crusts on other rock – resembling the magical interiors of limestone caves.
- Its colour varies, but it usually occurs in pastel shades of blue, pink, green and grey. Pink crystals come from Namibia, blue-green from New Mexico, and banded stalactite from Sardinia.
- Smithsonite used to be known as calamine (as was hemimorphite).

Similar stones

Chalcedony is often found in similar colours and forms, but is much harder.

Calcite crystals have a similar lustre and hardness, but calcite shows clear planes of cleavage.

Hemimorphite is lighter in weight and looks more fibrous and silky.

Magic

- Provides security in new relationships
- Instills confidence in times of flux

Healing functions

- Energizes the immune system
- Stabilizes emotions and releases stress
- Helps us to let go, find our own path and welcome new states of being
- De-clutters the mind, helping us to focus clearly

Practical ideas

- Wear smithsonite to ease anxieties about a new relationship.
- To prepare for important meetings: place smithsonite at the throat chakra or on the upper chest (at the thymus gland); a crystal of rose quartz at the heart; and amethyst at the solar plexus chakra.

Keywords

Blending

Combining

Smoothing

Blue-green smithsonite from New Mexico

Blue lace agate (SiO_2)

Agates can naturally take on all sorts of colour, but blue lace agate is one of the most subtle and delightful varieties. Its delicate, lacy bands of blue, white and grey create the impression of gentleness, calm and peace. In fact, its internal patterns are so beautiful that they can resemble distant, cloudy landscapes. These coloured layers, comprising microcrystals of quartz, are what define this type of chalcedony as an agate. It forms in geodes within volcanic rock, and South Africa has fine deposits. Blue lace agate is one of the more uncommon, and therefore more expensive, varieties of agate.

Colour: bands of blue, white and grey

Lustre: vitreous to greasy

Hardness: 6.5

System: trigonal, microcrystalline

Identification and care

- Blue lace agate is generally found as tumbled stones, slices and geodes.
- You may occasionally see it with some small terminated crystals if the central cavity of the geode allowed their growth.
- It is characterized by soft violet-blues appearing in delicate bands.

A natural piece of blue lace agate

Magic

- Brings peace, understanding and contentment
- Allows exploration of subtle planes of existence
- Conveys hidden messages

Healing functions

- Cools and calms any areas where there is a build up of energy in the body
- Gently nurtures and soothes the emotions
- Settles an overactive mind
- Engenders feelings of belonging in the world

Practical ideas

- Put a large tumbled stone of blue lace agate near your bed to encourage restfulness.
- To relax your eyes and calm your mind, gaze at the landscape-like patterns in a cut and polished slice of blue lace agate.
- To ease you through changes: place blue lace agate at the throat; amethyst at the brow or crown; rose quartz at the heart and solar plexus; and smoky quartz at the feet.

Keywords

Gently cooling

Nurturing

Comforting

Similar stones

Blue quartz does not have bands of colour.

Botswana agate has bands of greys, pinks, creams and browns.

Also: **chrysocolla** (see below)

171

Celestite (SrSO$_4$)

Also called celestine, celestite is a soft grey or blue mineral which has long been used in the making of ceramics, glass, flares and fireworks. In India, celestite powder is flung into sacred fires to create a spectacular deep red flame, which derives from the strontium it contains. Celestite often forms at very high temperatures in granite and pegmatite, but also crystallizes from the evaporation of salt water. Then it can be found alongside aragonite, chalk and sulphur. Wonderful sky-blue examples are found in Italy, Germany and Madagascar.

Colour: clear, grey-blue or sky-blue

Lustre: vitreous to pearly

Hardness: 3–3.5

System: orthorhombic

Identification and care

- Celestite is heavy for its size.
- Clusters are more common than individual crystals, which are stubby and wedge-shaped.
- Celestite is easily scratched; tumbled stones are more sturdy.

Magic

- Reveals new viewpoints
- Uncovers the fundamental energies of reality

A celestite cluster

four-sided, rhombohedral crystals

Healing functions

- Encourages communication and expression
- Helps to lighten our mood when needed
- Opens the mind to new ideas
- Encourages blissful silence, inspiration, meditative states and heightened intuition

Practical ideas

- For inspiration and a respite from worldly problems, place seven small clusters of celestite around the body: one above the head, one by each shoulder, one by each hand and one by each ankle for 10 to 15 minutes. Amplify the effects by lying on a white cloth, and keep grounding stones nearby to bring you back to reality.

Similar stones

Aqua aura (a modified quartz) has a six-sided cross-section, whereas celestite is four-sided.

Blue topaz is a much harder stone and usually has striations.

Blue quartz (SiO$_2$)

This stone combines the balanced clarity of quartz with the cool peacefulness of blue. Most blue quartz is massive (a mass of micro-crystals). In this form its colour is caused by inclusions of rutile or crocidolite, which bend the light passing through the quartz toward the blue part of the spectrum. Occasionally, large terminated crystals of transparent blue are found. Blue quartz tends to come from the USA, Brazil and Austria.

Colour: dense mid-blue, grey-blue, lavender-blue

Lustre: vitreous

Hardness: 6–7

System: trigonal

Identification and care

- In its massive form, blue quartz is almost indistinguishable from blue aventurine, although the latter has more "sparkle".

- Crystals of blue quartz are a transparent to translucent blue. They can be confused with the artificially coloured quartz aqua aura (see page 274), which displays a rich, almost metalllic, shiny blue on all surfaces, with no variations.

A blue quartz crystal cluster on a sandy matrix

174

Magic

- Calms agitated states
- Protects from negative influences
- Helps us to transmit information effectively

Healing functions

- Supports the heart, lungs, neck and throat
- Stimulates the immune system
- Brings contentment and a sense of hope for the future
- Dislodges deeply entrenched blocks to good communication
- Breaks set patterns of behaviour

Practical ideas

- Wear a blue quartz pendant to strengthen the immune system and improve communication skills and self-expression.
- To ease fears and sorrow, place blue quartz at the forehead and rose quartz at the heart.
- To activate personal spiritual potential, place three blue quartz on the body: one at the forehead and one in each hand.

Keywords

Protection

Contentment

Communication

Similar stones

Celestite has crystals that are four-sided, rather than six-sided.

Angelite is a softer stone with a softer colour.

Dumortierite is blue with areas of dark blue, brown or orange.

Sapphire has a more intense colour, with dichroism (its colour changes depending on the angle from which it is viewed).

175

Angelite (CaSO$_4$)

Also called anhydrite, angelite is quite a common sedimentary mineral that forms massive layers of rock. It is created when gypsum loses water, which causes the rock layer to shrink and caverns to form. It also occurs as a result of evaporation of sea water or the weathering of pyrite. The best angelite comes from Peru and Mexico, but it is also found in Germany.

The heart of angelite's healing action is the concentration of peacefulness. This crystal slows everyday thought processes and allows you to notice more subtle qualities of experience.

It is sometimes – mistakenly – thought that angelite takes its name from its angelic powdery blue colour, but it is actually named after Angela, the partner of the person who discovered the stone.

Identification and care

- Anhydrite forms gypsum-like crystals but is only called angelite in its massive blue form.
- Some samples fluoresce when exposed to ultraviolet light.

Colour: light blue, white, grey, violet

Lustre: vitreous to pearly

Hardness: 3.5

System: orthorhombic

Similar stones

Celestite is heavier than angelite, with transparent blue crystals.

Blue calcite is harder than angelite.

Blue topaz has a more glass-like quality.

Blue quartz is harder and usually opaque, rather than transparent or translucent.

- Angelite is soft and sensitive to water, so its polished surfaces should be treated with care to preserve the quality of their colour.

Keywords

Steadiness

Coolness

Attentiveness

Magic
- Promotes peaceful silence
- Enhances the ability to hear the subtlest communications
- Reveals hidden aspects of reality

Healing functions
- Eases constrictions of the throat and stiffness in the neck
- Soothes confused emotions
- Encourages creative communication
- Opens us to subtle communion with the finer layers of creation

Practical ideas
- For a temporary respite from over-excitement and confusion, hold angelite in your hand for a few minutes, followed by clear quartz, followed by a gently energizing stone such as carnelian or red jasper.

An anhydrite sample

177

Apatite (Ca$_5$(PO$_4$)$_3$(F.Cl.OH))

This crystal appears in many colours and forms, and often resembles other crystals: its name comes from the Greek word *apate*, meaning "deceiver". In fact, apatite was only recognized as a distinct mineral in the 18th century. It is present in most rocks but is most significant in sedimentary deposits where it may have formed from the remains of sea organisms. It is also the main component of human tooth enamel. Particularly rich sources are in Canada and Spain. In terms of healing qualities, apatite is linked to versatility and volatility.

Colour: blue, yellow or green

Lustre: vitreous to dull

Hardness: 5

System: hexagonal

Identification and care
- The most common examples are blue and green (sometimes called asparagus stone).
- Crystals have a "crackled" appearance beneath smooth surfaces.
- Apatite can look quite dull despite each crystal's variable shades of colour.
- Apatite is brittle and easily scratched.

Magic
- Transforms static situations

An apatite crystal

178

Healing functions

- Strengthens the physical structure of bones, teeth and muscle tissue
- Helps with personal expression and the flow of communication
- Encourages a flexible attitude
- Activates the intellect and deepens understanding
- Encourages subtle levels of awareness and perception
- Helps us to create structure
- Helps us to re-organize aspects of our lives that need attention

Practical ideas

- For a flexibile attitude, take three apatite crystals: place one at the throat chakra; and the other two either side of the sternum, below the collar bones.
- To optimize energy, take eight apatites: place one each side of the head, near the ears; one at the throat; one on the middle of the back; one on each side by the kidneys (level with the elbows); and one near each thumb.

Keywords

Structure

Flexibility

Strength

Similar stones

Kyanite is a similar blue, but its crystals are more blade-like and are rarely translucent.

Beryl has the same hexagonal form but no crackled appearance.

Topaz, unlike apatite, has a clear striation along its side and wedge-shaped terminations.

179

Tanzanite (Ca$_2$Al$_3$(SiO$_4$)$_3$(OH))

This crystal was only discovered in 1967 in the Merelani Hills near Mount Kilimanjaro in Tanzania. New precious stones always attract interest, especially when they have an ethereal quality suggestive of spirituality as this one does. Tanzanite is an example of the extraordinary emerging from the ordinary: it is a rare variety of the common mineral zoisite coloured by a few extra atoms within its lattice structure. It resembles sapphire but has a more violet hue caused by atoms of strontium. Its pleochroism means that three different colours – dark blue, brown-yellow and red-purple – can be seen in the same stone, depending on the angle from which you view it. Its blue colour can be deepened by heating.

Colour: strong blue with pleochroism of red, blue and brown

Lustre: vitreous

Hardness: 6–6.5

System: orthorhombic

Identification and care

- Gem-quality tanzanite is expensive and hard to find.
- Small chips or low-quality pieces are easier to find.
- The rose-pink variety of zoisite is called thulite (see pages 220–221).

A raw tanzanite sample

- Tanzanite should be handled carefully, as it is a brittle crystal.

Magic
- Reveals the extraordinary within the ordinary
- Associated with transmutation and revelation
- Can disguise the true nature of things

Healing functions
- Focuses healing energy wherever it is placed, but especially in the throat area
- Quickly releases strong pent-up emotions
- Stimulates intuition and psychic skills
- Integrates energy and information from diverse sources to create a new wholeness

Practical ideas
- To activate and amplify the energy of any stone, combine it with tanzanite.
- To experience increased spiritual perception, place tanzanite at the crown or brow, and a dark, grounding stone by the feet.

Keywords

Enlivening

Transmuting

Expanding

Similar stones

Sapphire rarely has the violet-purple tinges seen in tanzanite.

Amethyst generally occurs in darker shades of blue than tanzanite.

181

Sodalite $(Na_4Al_3Si_3O_{12}Cl)$

This is generally seen as a rich, royal blue mineral, shot through with white. However, it is allochromatic, meaning that each piece takes its colour from the impurities it contains. It therefore also exists in blue, green, yellow, pink, white and grey. Where sodalite exists alongside lazurite, it forms a significant component of lapis lazuli (see pages 186–187). Sodalite is found in igneous rock, particularly in Bolivia, Brazil, Equador and Greenland. In healing, the stone is particularly effective at quietening and stabilizing, encouraging calmness on every level of being.

Colour: usually blue with white veining

Lustre: vitreous to greasy

Hardness: 5.5–6

System: cubic

Identification and care

- Sodalite does form dodecahedral crystals but is generally seen in its massive form.
- When polished, good-quality sodalite can be difficult to tell apart from the more valuable lapis lazuli.

A polished sodalite stone

Magic
- Improves contact and communication with others
- Can help the transmission of messages over vast distances

Healing functions
- Cleanses the lymphatic system, enhancing the immune system
- Stabilizes emotions
- Clarifies perception
- Expands awareness during meditation
- Encourages peace and contentment

Practical ideas
- To release subconscious blocks, place sodalite at the brow, throat and sacral chakras.
- To relieve fears and inner conflict, place a Herkimer diamond at the brow, sodalite at the throat and carnelian at the sacral chakra.
- To ease a case of long-term discomfort, combine sodalite and malachite, either at the area of the stress or at the throat and heart chakras.

Keywords
Meditation

Contact

Peace

Similar stones

Lapis lazuli is a more intense, bright blue, and contains areas of white, rather than veining. Unlike sodalite, lapis lazuli contains pyrite.

Dumortierite is a duller violet-blue, with speckles.

Sapphire (Al₂O₃ (+ Fe and Ti))

Said to encourage magical workings, sapphire has long enchanted many collectors. Although generally considered a blue stone, this variety of corundum is found in a range of colours (red corundum, however, is ruby, see pages 86–87). After diamond, sapphire is the second hardest mineral known to man. Sri Lanka produces the most sapphires, although the finest come from Kashmir.

There is something solid, understated and reassuring about the colours and shapes of ruby and sapphire. They help us to focus upon lasting, worthwhile values. Such stones – the ones that our ancestors valued – are still often the best for restoring our equilibrium.

Colour: blue or violet-blue; many other colours, including black, green and yellow

Lustre: subadamantine to vitreous

Hardness: 9

System: trigonal

Identification and care

- Low-grade sapphires are available as tumbled stones or slices of hexagonal crystal.
- Clear, cornflower-blue sapphires are the most valuable; blue-black varieties are of less value.

184

Hexagonal sapphire crystals

- Star sapphires have rutile inclusions that create a star-light effect (asterism).

Protection

Understanding

Power

Magic
- Provides spiritual strength
- Protects from envy and harm
- Attracts favourable spirits and blessings
- Encourages wisdom and prophecy
- Aids victory over all (star sapphires)

Healing functions
- Balances the endocrine system and reins in overactive energies
- Calms, regulates and reduces tension in the solar plexus created by fear and anxiety
- Increases powers of personal expression, benefiting the heart and throat chakras
- Stimulates the higher mind

Similar stones

Apatite has more of a "crackled", translucent structure.

Ruby is similar in its uncut form, but shows a different colour.

Blue and clear topaz are similar when cut.

Beryl crystals do not narrow at the ends.

Practical ideas
- To encourage spiritual development, place six sapphires on and around the body: on the brow, solar plexus, palm of each hand and below each foot.

Lapis lazuli $((Na,Ca)_8(Al.Si)_{12}O_{24}(S,SO_4))$

This stone has a history that stretches back over 5,000 years. It was prized by the pharaohs of ancient Egypt, whose craftsmen used it to make treasures such as scarabs and eye-shaped amulets. Lapis was thought to be able to cure melancholy and recurring fevers, as well as increase mental clarity and intuition. It is likely that the ancient Egyptians got their lapis from the same mining areas that are in use today: Afghanistan has the best sources. A rock rather than a crystal, lapis lazuli's main component is blue lazurite. It sometimes includes sodalite, usually pyrite, and often has a matrix of calcite.

Colour: deep blues with white and gold flecks

Lustre: vitreous to greasy

Hardness: 5.5

System: cubic

Identification and care

- The finest examples are an intense blue, with small gold flecks of pyrite.
- Genuine lapis lazuli has a whitish glow when under ultraviolet light.
- Imitations are sometimes made out of glass with metallic flecks.

A raw lapis lazuli crystal

- Lower grades of lapis are sometimes crushed, rebonded with plastic or resin, and dyed to be sold at a higher price.

Magic
- Associated with truth, balance and justice
- Promotes steadfastness and courage

Healing functions
- Supports the throat and upper-chest areas
- Draws anxieties to the surface
- Aids communication and meditation
- Provides insight and clarity of mind

Practical ideas
- Place lapis at the forehead to encourage deep insight and memory recall.
- For discernment, truth and clarity, gather seven lapis lazuli crystals and five clear quartz ones. Place the lapis above the head, beside each shoulder, and by the hands and feet. Place the quartz near each ear, either side of the solar plexus and between the feet. Amplify the effect by lying on a blue cloth.

Keywords
Truth

Memory

Awareness

Similar stones

Sodalite has veining and is a less intense blue.

Dumortierite is a duller blue, with dark speckles, as well as pink and brown areas.

"Swiss lapis" is actually jasper stained blue.

Azurite $(Cu_3(CO_3)_2(OH)_2)$

A popular crystal for its beautiful deep blue colour, this stone has a startling blue edge, which is the key to its healing powers. Azurite forms when copper ores oxidize. It is associated with the green mineral malachite (see pages 152–153), with which it shares a similar chemical structure. In the Middle Ages, powdered azurite was used as a blue pigment in paints and dyes. As it absorbed atmospheric moisture, it used to oxidize to malachite, which explains why the sky is green in some early Renaissance paintings. The two minerals can be found in a form called azurite-malachite, which is excellent for "clearing out" deep-rooted emotions. The finest large azurite crystals come from Namibia.

Colour: blue to dark blue

Lustre: vitreous, chalky

Hardness: 3.5–4

System: monoclinic

Identification and care

- Small crystals appear a powdery blue and are often small, round nodules.
- Large crystals are often very dark and shiny, with an electric-blue edge.
- Crystals are soft, so need careful handling.
- Do not clean nodules in water.

A natural azurite sample from China

small needles and blades of azurite

Keywords

Deep release

Old memory

Integration

Magic
- Helps you to achieve swift results
- Reveals mysteries
- Transmits messages from the unknown

Healing functions
- Accesses deep levels of body-consciousness
- Draws out memories or old stress, allowing them to be released in healing
- Improves communication and creative flow

Practical ideas
- To encourage expansion of consciousness, place four pieces of azurite on and around the body: one above the crown, one either side of the head and one at the solar plexus.

Similar stones

Sapphire, unless of gem quality, is a less intense blue. It is also more glass-like and translucent.

Azurite-malachite stones have distinct areas of green and blue.

189

Dumortierite $(Al_7(BO_3)(SiO_4)_3O_3)$

An unusual and attractive stone, dumortierite has a variegated colouring that suggests an easy movement between disparate conditions and states. It is found in metamorphic rocks rich in aluminium and in some pegmatites, and is often intergrown with quartz. The best-quality stones come from Nevada, USA. This gemstone rarely forms large crystals, so is cut into cabochons and ornaments. Dumortierite will not melt (it is used in industry for lining furnaces): in healing, the stone's steadfastness helps us to maintain our equilibrium.

Colour: light blue to dark blue to violet, usually with dark speckles and pink and brown areas

Lustre: silky

Hardness: 7

System: orthorhombic

Identification and care

- Crystals form flat, blade-like, radiating clusters, but the mineral is usually sold in its tumbled form.
- Dumortierite closely resembles other blue stones, but can usually be distinguished by its speckles and its rich range of blue, brown and violet tones.

Tumbled dumortierite

- It has a more earthy, everyday feel than some other dark blue stones such as sodalite, lapis lazuli or azurite.

Magic
- Resolves opposing points of view

Healing functions
- Slows down aggravated and irritated energies
- Cultivates patience
- Creates detachment, allowing us to maintain personal equilibrium
- Helps us to express spiritual ideas and understand hidden meanings
- Stimulates communication between the body's various systems
- Promotes subtle understanding and easy flow

Practical ideas
- Dumortierite will help to bring calm without lowering energy levels too much: carry a stone to restore easy flow to everyday life.
- Place dumortierite at the throat chakra to encourage the sharing of advice with others.

Keywords

Understanding

Conciliation

Equilibrium

Similar stones

Blue quartz and **blue aventurine** both have a more granular appearance than dumortierite, with white or metallic elements.

Kyanite has similar blade-like crystals, but a more shiny lustre.

Lapis lazuli is a more vibrant blue, with white and gold pyrite areas.

Sodalite is deeper violet-blue with white veining.

191

Kyanite (Al₂SiO₅)

Forming in thin, blade-like crystals that often radiate from a common centre, kyanite is found in metamorphic rock. Most samples come from Austria, Brazil, France, Italy and Russia. The white or grey variety is called rhaeticite. Kyanite is unusual in that its hardness depends on the direction in which a crystal is stressed (because of this, the stone is also known as disthene, meaning "two strengths"). It is rarely of gem quality but is useful in industry for its resistance to heat and low electrical conductivity. In healing, however, kyanite is an excellent conductor of life-energy.

Colour: usually blue to blue-black, sometimes streaked; also white, grey or green

Lustre: mother of pearl, vitreous

Hardness: 4–7 (depending on direction of stress)

System: triclinic

Identification and care

- Kyanite is usually seen in clusters of thin blades, which are glassy but opaque.
- Gem-quality slices or cabochons resemble apatite or sapphire.
- Parallel striations with cross-grains are visible along the length of the crystal.

A classic kyanite cluster

- Delicate clusters need careful storage; thicker single crystals are quite sturdy.

Magic
- Acts as a catalyst to speed up other processes
- Makes rapid links between all things
- Aids rapid communication

Healing functions
- Balances the whole being
- Calms and releases blocked emotions
- Brings tranquillity and quietness
- Aids meditation and access to all levels of awareness

Practical ideas
- Place kyanite on any area of the body that feels over- or under-energized.
- Place a crystal on the forehead to quieten the mind and encourage clear thought.
- To stop recurring worries, place a kyanite crystal at the brow, one on each wrist pulse and a smoky quartz crystal at the feet.

Keywords

Connecting

Frictionless flow

Release

Similar stones

Apatite has a more "crackled", translucent structure.

Indicolite, when of gem quality, is similar in colour, but not as flat.

Sapphire crystals are harder and hexagonal.

Sillimanite (fibrolite) has an identical composition, but is orthorhombic (not triclinic), with white, needle-like crystals.

193

Preseli bluestone (composition varies)

This stone, some 480 million years old, is a form of dolerite found only in the Preseli Mountains of Pembrokeshire in Wales. Dolerite is a rock composed mainly of calcium feldspar and augite with traces of copper and pyrite. Five thousand years ago, the Neolithic people of Britain moved 19 giant pieces of bluestone, each weighing up to four tons, from Wales, more than 200 miles, to Salisbury Plain, to build the inner circle at Stonehenge. Legend has it that Merlin the wizard used his magic to move the bluestones to Stonehenge through the air. Preseli bluestone has a powerful aura, giving focus and stability and anchoring us in the Earth's energy.

Colour: blue-grey, blue-green with inclusions	
Lustre: dull, unless polished	
Hardness: variable	
System: none	

Identification and care

- Preseli bluestone looks blue-grey when unpolished and blue-green when polished.
- In polished pieces, grey inclusions of feldspar give the impression of stars in the night sky.

A piece of natural, unpolished Preseli bluestone

194

Magic

- Connects us with the distant past
- Reinforces the effect of ritual use of sound and movement
- Provides a doorway to different dimensions

Healing functions

- Simultaneously relaxes and energizes
- Focuses healing to the ears, nose and throat
- Cuts through emotional entanglements, promoting grounded contentment
- Steadies the mind, encouraging increased clarity of thought

Practical ideas

- To access the deep mind and far memory, place one bluestone above the brow chakra, near the hairline, and put two more bluestones or other blue crystals or clear quartz next to the elbows. Rest your hands on your solar plexus.
- To bring deep calm, put one bluestone on the brow and another at the base of the skull (occiput).

Keywords

Clarifying

Grounding

Timeless

Similar stones

Basalt has a less granular, more even texture, with visible feldspar and olivine crystals.

Granite is more granular, with large crystals. It is rarely dark in colour.

Seraphinite has a similar play of light, but is softer.

Dumortierite is a lighter blue and is speckled.

Blue quartz and **blue aventurine** are denser and a more even blue.

195

Amethyst (SiO$_2$)

With its stunning purple colour, amethyst has been a gem favoured by royalty throughout the ages. Its name comes from the Greek for "not drunken", and, since classical times, amethyst has been believed to moderate unruly behaviour and to encourage clarity of mind and spirituality. It is a good all-round healer.

Amethyst is a variety of quartz coloured by iron. Shades vary from pale lilac to dark purple, but it can turn brown, yellow or green if exposed to heat. The crystal mostly forms in geodes, in rock cavities left by volcanic gas bubbles.

Colour: violet	
Lustre: vitreous	
Hardness: 7	
System: trigonal	

Identification and care

- Amethyst shows dichroism, appearing blue-purple or red-purple from different angles.
- The colour tends to be strong at the tip and milky at the base.
- Most amethyst is found in short-sided prismatic crystals.
- Mexican amethyst tends to be pale with long, narrow prisms.
- Chevron amethyst displays bands of purple and white.

Similar stones

Fluorite resembles amethyst and chevron amethyst in its violet and "blue John" varieties respectively. However, both these fluorite forms are cubic and softer.

Ametrine is a blend of violet and yellow quartz, created by heat naturally or artificially.

Magic

- Helps to maintain equilibrium in all situations
- Balances polarities

Healing functions

- Encourages self-control
- Calms harsh emotions and brings stability
- Helps practical applications of imagination
- Aids meditation and sleep

Practical ideas

- To promote healing, take eight amethyst crystals: place one above the head, one below the feet and the rest evenly around the body (all pointing inward). Amplify the effects by lying on a yellow or violet cloth.
- This healing layout also creates a safe space in which to explore subtle perception. To explore the qualities of other crystals, place them on the brow, in turn, finishing with a grounding stone, such as black tourmaline.

An amethyst crystal cluster

197

Fluorite (CaF$_2$)

Although this mineral comes in many colours, rich purple is the most common. It was the first mineral to show fluorescent under certain lighting and gave its name to that phenomenon. It also forms fascinating shapes: interpenetrating, cubic crystals often coated with other minerals, such as calcite, quartz or pyrite. Fluorite is an important component in many industrial processes, such as glass- and steel-making. Notable sources are the UK, Germany, Italy, Switzerland and the USA.

Colour: purple, blue, green, yellow or clear

Lustre: vitreous

Hardness: 4

System: cubic

Identification and care

- Fluorite crystals are found in stepped cubes that look like miniature cities.
- Single octahedral crystals occur naturally, but most examples on sale are artificial.

Magic

- Encourages innovation and invention
- Helps in planning and coordinating resources

A fluorite crystal cluster

cubic crystal structure

Healing functions

- Supports healthy bone tissue and physical structures of all organs
- Helps us to master physical skills and improve dexterity and balance
- Encourages a sense of self-worth
- Facilitates assimilation of new ideas and understanding of fine levels of awareness

Practical ideas

- Place a large fluorite crystal or cluster in a study or workplace to help orderliness.
- Place fluorite beside the bed to help you to find new ideas and solutions while you sleep.
- Carry fluorite somewhere on your upper body to improve coordination skills.

Similar stones

Amethyst has no internal cleavage and is harder than fluorite.

Halite is similar in crystal form, but is softer, lacks lustre and is not transparent.

Galena is similar in crystal form, but has a dull metallic silver sheen, is heavy and is toxic.

Calcite has a similar appearance but tends to be single-coloured, unlike fluorite.

199

Sugilite $(KNa_2(Fe_{2+},Mn_{2+},Al)_2Li_3Si_{12}O_{30})$

Sugilite is formed in the volcanic rock syenite. It was first found in the early 20th century in Japan. Attractive new minerals always excite enthusiastic interest and as other deposits came to light elsewhere, particularly in South Africa, it acquired other trade names, such as Royal Azel, lavulite and luvulite. Sugilite is usually found in a granular, massive form with deep purple colouring and darker veins, making it excellent for cutting and polishing. Purple is often considered a magical and spiritual colour; sugilite conveys the attributes of purple within a massive mineral form, suggesting an earthy, balanced approach to subtle experience.

Colour: pink, lilac or purple	
Lustre: vitreous to resinous	
Hardness: 5.5–6.5	
System: hexagonal	

Identification and care

- Sugilite varies markedly in colour from bright purple to almost black.
- It is keenly sought after, which means that it is prone to imitation, especially in the form of dyed howlite.

A piece of unworked sugilite

Magic

- Helps us to achieve spiritual goals
- Integrates the subtle and the material

Healing functions

- Coordinates the left and right hemispheres of the brain and encourages energy flow within nerve cells, helping with learning difficulties and physical dexterity
- Reduces tension between what is desired and what can practically be achieved
- Integrates sensitive individuals with their everyday surroundings
- Gives a practical focus for spiritual energies

Practical ideas

- To feel at ease with your spiritual nature, place a piece of sugilite at the brow, turquoise at the throat chakra and tiger's eye at the sacral chakra.
- To counteract feelings of alienation and isolation, place clear quartz at the crown, sugilite at the throat, moldavite at the heart and smoky quartz at the feet.

Keywords

Integration

Sensitivity

Spiritual confidence

Similar stones

Purpurite has a softer, more fibrous surface than sugilite.

Charoite includes different shades in the same stone, often in sinuous patterns.

Dyed howlite tends to be a brighter, more even colour than sugilite.

Charoite $((K,Sr,Ba)(Ca,Na)_2(Si,Al)_4O_{10}(OH,F))$

Charoite originates from the region of the River Chara in Siberia where it is found in alkaline igneous rocks. When polished, it shows swirling purple and white cloud-like structures interspersed with translucent patches and inclusions of black aegerine (a pyroxene). The essence of charoite is movement and flow, encouraging harmony on a deep level. As with most other purple stones, it helps to integrate different aspects of our lives by combining the energies of red (practicality, action and involvement) and of blue (quietness, introversion, peace and detachment).

Colour: violet-purple with white, pink and gold	
Lustre: vitreous to dull	
Hardness: 5–6	
System: triclinic	

Identification and care
- Charoite is highly decorative and is often cut as cabochons, beads and slabs.
- Its sinuous fibres make it easy to distinguish from other opaque purple crystals.

Magic
- Encourages understanding
- Helps us to accept change
- Brings balance amid turbulence

Similar stones

Sugilite is granular and massive, and does not have crystalline inclusions.

cloud-like marbling

Keywords

Calm perspective

Flexibility

Awareness

Healing functions

- Soothes tense emotions
- Calms the nervous system
- Encourages a personal perspective on collective situations

Practical ideas

- To help you adapt to a new situation, place charoite at the heart chakra, surrounded by green tourmalines, and place a grounding stone by the feet.
- To quieten and soothe the nervous system, place charoite at the crown chakra and moonstones at the throat and solar plexus chakras.

A polished slab of charoite   203

Iolite (Mg$_2$Al$_3$(AlSi$_5$O$_{18}$))

Also known as "water sapphire", iolite is similar in colour to sapphire – a translucent, violet-blue – but from some angles it shows brownish-yellow or pale, blue-grey. It is fairly common and found in sedimentary gneisses. The best sources of iolite are Sri Lanka and Myanmar, which produce river-tumbled stones. As a crystal that changes colour depending on the angle from which you view it, iolite has the ability to reveal new perspectives.

Colour: violet, violet-blue, brown or blue-grey

Lustre: greasy to vitreous

Hardness: 7

System: orthorhombic

Identification and care

- Iolite is often massive, granular or in short columnar crystals.
- Inclusions of haematite turn some varieties of iolite red.
- A rough iolite stone is granular and partially translucent.
- Grey varieties can resemble jade and are used in carving.

Magic

- Enables us to consider a situation from all angles

A rough piece of iolite

- Combines science and magic, logic and imagination

Healing functions
- Activates the nervous and endocrine systems
- Enhances feelings of sympathy and empathy – both in us for others and in others for us
- Helps us to decide when other people do or do not need our help
- Encourages understanding and acceptance of the duality of existence

Practical ideas
- When you have to make a difficult choice, use three iolites, placed on the brow and on each side of the head (or held in each hand).
- If you need to resolve a difficulty within a relationship or decide how to deal with strong feelings toward someone, place an iolite at the heart, another at the solar plexus and a garnet between the feet.

Keywords

Clear thinking

Empathy

Decision-making

Similar stones

Blue topaz shows no internal changes of colour.

Indicolite (blue tourmaline) shows only slight dichroism (an internal change of colour).

Sapphire (see below) has a metallic sheen, whereas iolite is more grey-blue.

Purpurite $((Mn^{3+},Fe^{3+})(PO_4))$

Hardly looking like mineral at all, the soft, silky sheen of purpurite's tightly arranged, fibrous crystals makes it easy to identify. It is somewhat rare (principal sources are in Rwanda, Namibia and the USA), but its decorative quality has given it a place in the crystal collector's market.

Purpurite can be a useful stone for healing, particularly for easing pain and for repairing physical or emotional damage in a sympathetic and gentle manner. The stone's silkiness encourages subtle energies to flow smoothly.

Colour: dark violet to dark red

Lustre: dull to velvet

Hardness: 4.5

System: orthorhombic

Identification and care
- Purpurite appears to have a soft and silky texture.
- It is sometimes cut into cabochons to show off its rich colour and pleasing texture.

Magic
- Helps to integrate disparate skills, thoughts and ideas
- Helps us to repair mistakes

206

A natural piece of purpurite

Healing functions

- Maximizes nutrient intake in the body, helping the generation of energy
- Relaxes and soothes the mind
- Strengthens intuition
- Encourages acceptance
- Provides an underlying support to all life-energy
- Pulls disparate strands of our persona together helping us to achieve spiritual goals

Practical ideas

- To create harmony in mind, body and spirit, place a piece of purpurite at the crown or brow chakra, one at the heart and one at or near the root chakra.
- To ease stress and frustration when you are feeling a lack of direction or progress, place a piece of purpurite at the brow and a piece of black tourmaline between the feet.

Keywords
Contentment
Acceptance
Focus
Calm

Similar stones

Sugilite has a shinier, less fibrous surface than purpurite.

Rose quartz (SiO$_2$)

This crystal, one of the most valued quartz varieties, is renowned for its delicate pink tone. The impurities of titanium or manganese that give it this hue also prevent the growth of large individual crystals, which means it is usually found in massive form. Most rose quartz these days comes from Brazil but there also sources in India, Madagascar and the USA. In healing work, the stone encourages the heart to open, bringing more trust and love into our lives.

Colour: pink, rose, peach, violet pink

Lustre: vitreous

Hardness: 7

System: trigonal

Identification and care

- Large, single rose quartz points have been shaped from larger rough pieces.
- Unpolished rose quartz has sharp edges.
- "Lavender" quartz is rose quartz with a tinge of violet.
- Most pieces of rose quartz are translucent with many fractures; gemstones are flawless and transparent.

Massive rose quartz

- Star rose quartz containing rutile crystals has asterisms (star effects), which can be seen when light reflects off the polished surface.

Magic
- Fosters an appreciation of beauty
- Inspires feelings of love and friendship

Healing functions
- Rapidly releases emotional stress – the effect is so intense that it can be uncomfortable (to avoid this, balance with grounding stones)
- Uncovers the underlying causes of other problems, such as a negative self-image

Practical ideas
- To release emotional stress, place tiger's eye at the sacral chakra, grounding stones by the feet and rose quartz at the heart, surrounded by four clear quartz (points out).
- For self-worth and nurturing, place 12 rose quartz evenly around the body. The effect of the experience is amplified if you lie on a pink or magenta cloth.

Keywords
Love

Emotion

Release

Similar stones

Gem-quality rose quartz is very similar to **morganite**, **topaz** and **kunzite**.

209

Kunzite (LiAlSi$_2$O$_6$)

This beautiful pink crystal, named after the Amerian gemstone specialist G. F. Kunz, who discovered it at the start of the 20th century, is the most sought-after member of the spodumene family (to which the delicate green hiddenite also belongs). It forms long, striated prisms which can be opaque but are often deeply translucent or completely transparent. Kunzite's colour from a side view is pale, but looking down the length of the crystal, it deepens to an appealing pinky violet. Occasionally, spodumene crystals will contain areas of pink and green together.

Colour: pink, lilac pink (hiddenite is green)

Lustre: vitreous

Hardness: 6.5–7.5

System: monoclinic

Identification and care

- Kunzite crystals contain parallel striations.
- Stones darken in colour once mined.

Magic

- Strengthens the heart and attunes us to universal love
- Counteracts aggression
- Helps us to understand other people
- Brings conflict to a resolution

Tumbled kunzite

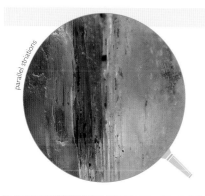

parallel striations

Keywords

Love

Support

Understanding

Peace

Healing functions

- Supports the cardiovascular system and thyroid function
- Enhances self-esteem
- Helps to override negative, unhelpful thought patterns – very protective
- Creates a space for meditation

Practical ideas

- To heal and protect where love is lacking, place kunzite at the heart and arrange four clear quartz, points outward, in a diamond shape around the body.
- Wear a pendant of kunzite or hiddenite to provide yourself with invaluable protection during difficult emotional times.

Similar stones

Rose quartz of a high quality is similar, but has no striations.

Morganite and pale **rubellite (pink tourmaline)** display no change in colour depth when you look along the central axis.

211

Lepidolite (K(Li,Al)$_3$(Si,Al)$_4$O$_{10}$(OH,F)$_2$)

This crystal, which has only become widely available over the past decade, can be found as masses of rounded sheaves or as small, flat intergrown plates, but most often you will see it as microcrystalline stone with a violet-pink colour and a surface that glistens as it catches the light. Lepidolite is a variety of mica, coloured by lithium. In industry mica acts as an insulator; in healing, lepidolite insulates the heart from hurt. Look for lepidolite intergrown with crystals of pink tourmaline (rubellite) – which happens quite commonly. This gives the soft lepidolite added strength and increases its ability to heal emotional pain and encourage the expression of desires.

Colour: pink, violet, grey	
Lustre: pearly	
Hardness: 2.5–3	
System: monoclinic	

Identification and care
- Tumbled lepidolite is a rich to pale pink, often with included pink tourmaline.
- Tumbled stones may not "glisten" much.

Magic
- Confuses enemies
- Makes true feelings known

Similar stones

Rubellite (pink tourmaline) can look similar when tumbled, but usually striations are visible.

Thulite has no "sparkle".

Healing functions

- Balances nerves and muscles, relieving excess tension carried in the body
- Brings a sense of inner security, safety and protection from harm
- Helps us to trust our instincts and to assess different viewpoints
- Encourages us quietly to go about reaching our full potential

Practical ideas

- For courage, strength and self-confidence, place lepidolite at the throat, thymus and heart, citrine quartz between the solar plexus and navel, and a garnet or star ruby by the feet.
- To bring a deep sense of relaxation, place a piece of lepidolite at the solar plexus and imagine your breath entering and leaving your body through the stone.

Keywords
Fulfilment
Security
Instinct

A "blade" of lepidolite

213

Morganite $(Be_3Al_2Si_6O_{18})$

Morganite's pink colour, which is caused by the presence of lithium, manganese or caesium, makes it one of the rarer varieties of beryl (emerald and aquamarine being the best known of the other coloured beryl varieties). Like all pink stones, it is good at resolving issues that relate to self-worth and self-identity. When worn as a pendant near the heart chakra, morganite may begin to uncover and dissolve deep emotional conflicts and painful memories. Such work is best done gradually to minimize any discomfort. Morganite also has a very protective energy, reducing aggression and encouraging tolerance around us.

Colour:	pink
Lustre:	vitreous
Hardness:	8
System:	hexagonal

Identification and care

- Tumbled stones can be almost transparently clear or an opaque milky white.
- Look for striations, which are common to all beryls.
- Morganite often takes a higher polish than quartz.

A hexagonal crystal of morganite

Magic
- Expresses perfect clarity, peace and openness
- Melts boundaries
- Makes everything seem possible

Healing functions
- Works at the sacral chakra to soothe and nurture the reproductive system
- Helps us to understand and resolve our conflicts with other people, especially where small irritations have grown into larger issues
- Encourages us to put the various aspects of our lives in perspective, establishing priorities
- Ideal as a focus for a meditation to link us to the energy behind creation

Practical ideas
- Place morganite at the sacral chakra to help with the resolution of sexual issues and emotional blocks. This will also help you to work through deep trauma and loss.
- Place morganite near any chakra to bring insight into the issues affecting that chakra and how to resolve them.

Keywords

Release

Simplification

Love

Similar stones

Rose quartz, rubellite, pink topaz and kunzite are all very hard to tell apart from morganite if they are cut or tumbled.

Rhodonite (MnSiO₃)

This pink stone is named after the Greek word for rose: *rhodon*. Where crystals occur, they are table-top shapes with rounded edges, but rhodonite is more often found in massive form. A good-quality polished slice can resemble the deep pink section of a ripe watermelon. Other pieces are more salmon pink, with veins of brown or black impurities. The pink colour links us with the underlying energy of creation, while the brown or black veining enables a practical use of these fine-level forces. Most rhodonite is found in metamorphic rocks.

Colour: pink to dark pink with brown or black areas

Lustre: vitreous

Hardness: 5.5–6.5

System: triclinic

Identification and care
● Gem-quality rhodonite has translucence, and crystalline areas.

Magic
● Makes deep needs manifest themselves
● Helps us to make ourselves heard, improving communication channels and facilitating mutual understanding
● Traditionally associated with sympathetic magic

A piece of natural rhodonite

Healing functions

- Supports us during times of transformation, especially if the change is not of our choosing
- Enables us to express strong, genuinely felt emotions
- Directs emotions toward goals
- Dispels negative states of mind, confusion and anxiety
- Increases the effectiveness and sensitivity of a meditation supported by a mantra

Practical ideas

- Rhodonite attunes us to subtle vibration, so hold or wear a piece for performing chants and mantras, particularly during meditation.
- To make yourself fully heard, place rhodonite at the sacral chakra and throat.
- To unblock communication channels, place at the base of the skull and throat.
- To develop clairaudience (subtle hearing), place a piece at the sacral chakra and a piece close to each ear.

Keywords

Resonance

Passion

Motivation

Communication

Similar stones

Rhodocrosite of gem quality can closely resemble the single-coloured crystals of rhodonite.

Strawberry quartz can look like low-quality rhodonite.

Thulite in its raw form resembles rhodonite, but has more sparkle.

217

Rhodocrosite (MnCO₃)

Often resembling a version of blue lace agate in reds, oranges and pinks, rhodocrosite owes its colour to the presence of manganese. It has concentric colour bands that appear when its botryoidal crystal form is cut and polished into slabs or cabochons. The deeper pink bands often contain small rhombohedral crystals. Large single-coloured crystals are an intense, transparent deep pink-red and are hard to find.

Some rhodocrosite brings to mind ripples on a pool covered with rose petals. Such a soothing image is in keeping with the stone's ability to relieve all areas of the body that are known for storing tension and stress, particularly the solar plexus and abdomen. The range of colour, from the most intense deep pink, to pale pink to orange and cream, helps to harmonize the root, sacral and solar plexus chakras.

Colour: pink, reddish-brown, grey

Lustre: vitreous

Hardness: 3.5–4.5

System: trigonal

Identification and care
● Crystals are rare, expensive and found in beautiful rhombohedral shapes.

A polished section of botryoidal rhodocrosite

- Tumbled stones, cabochons or slices are readily available for use in healing.

Healing functions
- Eases difficulties within the digestive and reproductive systems
- Enhances self-worth and self-confidence
- Releases tension caused by anxiety
- Relieves disturbances in the aura caused by trauma
- Drives away pain and anger

Practical ideas
- Placing a rhodocrosite stone on the lower abdomen will ease any discomfort or pain in this area, particularly any energy blockages that may be associated with doubt, fear or lack of self-confidence. You can amplify the effect by placing a Herkimer diamond at the sacral chakra.
- Place rhodocrosite at the pulse points on the wrists to strengthen your sense of self and enhance your ability to achieve goals.

Keywords
Self-confidence

Ease

Empowerment

Similar stones

Coral is a deeper orange-pink.

Rhodonite has brown or black veining.

Cherry opal (see below) has more areas of translucent white.

Thulite $(Ca_2Al_2OH(SiO_4)_3)$

Known as an ornamental carving stone in Norway since ancient times, thulite is a massive variety of the mineral zoisite, found in metamorphic limestones and also in igneous rocks where the original minerals have been altered. The stone has a delicate strawberry pink colour (caused by manganese), and may also contain the green more familiar with other zoisites. Named after the mythical island of Thule, the land beyond the North Wind, thulite is associated with the furthest limits of the world, instilling in us the confidence to go beyond our normal limits.

Colour: pink, red-pink, grey	
Lustre: vitreous to pearly	
Hardness: 6–6.5	
System: orthorhombic	

Identification and care

- Thulite occurs in granular aggregates and as polished cabochons and slices.
- It has an even colouring of mid-pink.
- Lower grades have areas of grey-white and may contain green zoisite as well.

Magic

- Gives us the confidence to explore and attracts us to the unknown

Similar stones

Rough pieces of **rhodonite** are similar in colour to thulite, but have dark veins.

Tumbled **strawberry quartz** is more vitreous.

Healing functions

- Enhances the performance of the nervous system, facilitating communication between the brain and the rest of the body
- Helps us to move on from unhelpful recurring emotional states
- Motivates us to take action when we feel complacent or unenthusiastic

Practical ideas

- To calm the nerves, place thulite at the solar plexus.
- To increase self-assurance, place thulite at the sacral chakra, green aventurine at the heart and blue lace agate at the throat.
- To welcome a new experience about which you feel apprehensive, lay out four thulites in a diamond shape around the body. If you focus on the situation that is making you anxious while lying in this pattern, it will help to allay your fears.

Keywords
Exploration
Self-assurance
Motivation

Unpolished thulite

221

Moonstone (KAlSi$_3$O$_8$)

Called adularia in mineralogical terms, it is very apt that this variety of the mineral feldspar is named moonstone, for it has a soft, pearly lustre, reminiscent of the moonlight. Large crystals are rare, and moonstone is usually found in microcrystalline massive form. The most important source of moonstone is Sri Lanka. Highly regarded in India as the perfect gemstone for women to wear, it is the archetypal stone of lunar energy and the feminine spirit, and enhances feminine virtues.

Colour: pearly white, cream, yellow, blue

Lustre: vitreous

Hardness: 6–6.5

System: monoclinic

Identification and care

- The characteristic lustre of moonstone can vary depending on structure. It can show opalescence, rainbow iridescence and schiller (twinkling), all caused by light reflecting off internal structures.

- Where there are alternating layers of albite and orthoclase, a blue colour or a milky-white schiller can appear.

- Low grades of moonstone can be grey, brown, yellow or orange.

A natural piece of moonstone

Magic
- Attunes us to the cycles of time
- Harnesses the power of the moon and water
- Conveys goddess energy
- Associated with fertility, flow and growth

Healing functions
- Balances the body's blood and lymph systems (aspects influenced by the element water)
- Relieves menstrual cramps and helps with other female issues
- Relieves indigestion, especially if stress-related
- Stabilizes the emotions and releases tension
- Helps us to empathize with other beings
- Enhances intuition and imagination

Practical ideas
- To soothe and calm the body, emotions and mind, use five moonstones, placing them: above the head, on each shoulder above the armpit and on each hip bone. This layout is amplified if you lie on a dark blue cloth.

Keywords
Emotional balance
Release
Empathy
Femininity

Similar stones

Other types of **feldspar**, such as **albite**, might resemble poorer grades of moonstone.

Spectrolite ("rainbow moonstone") is more similar to labradorite, but it is white with vivid areas of iridescence.

Zeolite (complex calcium, potassium and sodium silicates)

Members of the zeolite family often grow together on a common matrix and can be difficult to tell apart. Individual varieties are of limited interest to the crystal healer, but together they exude a strong collective energy, suggestive of harmony and cooperation.

Colour: white, grey-white, yellow, red

Lustre: vitreous to pearly

Hardness: 3–5

System: mostly monoclinic

Identification and care

Individual zeolites are described below:

• **Stilbite** (desmine) has a pearly lustre and a bow-tie sheaf of twin crystals.

• **Heulandite** forms tabular crystals of white or red.

• **Phillipsite** forms globular, radiating aggregates or small intertwined crystals of whitish, pearly lustre.

• **Chabazite** forms white or red rhombohedral or cube-like crystals.

• **Natrolite** is the most common zeolite, the often translucent needle-like crystals form spherical aggregates, clusters and coatings.

A heulandite crystal cluster

- Zeolite crystals are most often seen as display pieces in association with apophyllite and okenite.
- All zeolites are soft, so they need careful cleansing and should be stored separately.

Magic
- Encourages innovation, humour and support
- Motivates groups of people to stand together
- Provides access to collective consciousness

Healing functions
- Makes body systems work in harmony
- Instills in us the confidence and drive to begin new ventures
- Helps us to show sensitivity to others
- Suggests a broader, more optimistic approach

Practical ideas
- To build your awareness of the meaningful relationships between all sentient beings, place a zeolite (of any variety) at the crown, celestite at the throat, jade at the heart and lodestone at the feet.

Similar stones

Calcite is less pearly and has reflective surfaces.

Albite and other feldspars may have a similar appearance, but are harder and associated with igneous rocks and crystals.

225

Milky quartz (SiO$_2$)

Common in mountainous regions, veins of milky quartz can be seen threading their way across exposed and eroded rock surfaces. Reflecting light from great distances, this stone has been used to enhance sacred sites for thousands of years. The stones we buy and use today retain this aura of sanctity.

Milky quartz is pure white, either opaque or translucent, and owes its colour to tiny bubbles of air or water contained within the quartz. When other impurities, such as iron, slightly colour the crystal, it is known as common quartz. Areas of increased milkiness often occur toward the base of clear quartz crystals – again the result of tiny air or water bubbles. Sometimes these veils can clear when we hold the crystal, as our body heat drives out the bubbles through microscopic fractures.

Colour: white

Lustre: vitreous

Hardness: 7

System: trigonal

Identification and care

● Some coloured varieties of common quartz have distinct names – for example, the pinky-brown "strawberry quartz".

A terminated milky quartz crystal

- Milky quartz is usually massive, but fully grown crystals also occur.

Magic
- Symbolizes the energies of the moon
- Represents purification and cleansing
- Harnesses the power of light

Healing functions
- Clears blocked energy gradually
- Gently releases difficult emotions
- Disperses negative thoughts
- Calms an overactive mind
- Helps to ease us into the universal flow

Practical ideas
- To become better connected with yourself and the world, gather six milky quartz tumbles and six black tourmalines and place as follows: milky quartz beside the shoulders, by each hand, and beside each foot; tourmalines above the head, below the feet, by each elbow, and by each knee. For increased effect lie on a deep pink cloth.

Keywords
Diffusing
Soothing
Relaxing

Similar stones

Marble (a rock derived from limestone or dolomite, composed of calcite, calcium and magnesium silicates) tends to look opaque and more granular.

227

Selenite (CaSO$_4$.2H$_2$O)

Named after Selene, the Greek goddess of the moon, because of its luminous moon-like glow, selenite is a gem-quality form of the common mineral gypsum. Gypsum was formed by the evaporation of ancient seas and salt-water lakes, and selenite hints at its origins both in its watery-clear beauty and its extreme solubility.

Colour: white, clear (occasionally yellow, brown, red or blue)

Lustre: vitreous to pearly

Hardness: 2

System: monoclinic

Identification and care

- Selenite often forms twinned crystals with a characteristic "swallow-tail" shape.
- Fibrous, translucent white massive forms of selenite are called satin spar.
- Round, rosette-like concretions of selenite are called "desert roses".
- Selenite is very soft, easily scratched by a fingernail.
- It is extremely sensitive to water and humidity – it will bend even in the hand and may slide apart into many thin sheaves if it is put into water.

Two pieces of tumbled selenite

Magic
- Clears away darkness and shadows
- Helps us to avoid difficulties
- Protects us from harm and blame

Healing functions
- Quickly unblocks stagnant energy with its rapidly flowing and cooling energy
- Shifts negative emotions, especially resentment and long-standing anger
- Brings clarity of mind
- Expands consciousness

Practical ideas
- Holding a natural crystal wand of selenite can help to purge the day's tensions as well as painful memories that are difficult to release. Imagine breathing in through the top of your head, then breathing out along the wand. Visualize your troubles flowing out of the stone, neutralized and harmless.
- Alternatively, place selenite at the sacral chakra to remove stress, trauma and stubborn imbalances in the body.

Keywords
Expansion
Release
Cooling

Similar stones

Ulexite (see below), also known as TV stone, has striated surfaces similar to selenite but clearly carries light and images along its fibres.

Tourmaline quartz

(SiO$_2$ with Na(Mg,Fe,Li,Mn,Al)$_3$Al$_6$(BO$_3$)Si$_6$O$_{18}$(OH,F)$_4$)

Quartz and tourmaline often crystallize together
to form tourmaline quartz: quartz that has long,
thin needles of tourmaline embedded within
it. It usually contains the most common variety
of tourmaline – the black crystals of schorl –
but it can be any kind. Tourmaline quartz
brings us both the enlivening energy of quartz
and the grounding energy of tourmaline –
a powerful combination.

Colour: white or cloudy
with black linear inclusions

Lustre: vitreous

Hardness: 7

System: trigonal

Identification and care
- Tourmaline quartz is found as massive
  lumps, included (or intergrown) crystals,
  tumbled stones, beads and cabochons.
- The quartz can be clear or cloudy.

Magic
- Offers protection from
  harmful energies
- Encourages clear, well
  thought-out, decisive action
- Helps us to embrace Earth energies

A cut sphere of tourmaline
quartz from Brazil

thin needles of tourmaline

Healing functions

- Strengthens all levels of body functioning
- Deflects environmental disturbances
- Keeps our thoughts and emotions anchored in the present
- Helps us to sift out unwanted thoughts
- Integrates energy patterns, aligning all subtle bodies and chakras, and correcting energy levels throughout

Practical ideas

- Place tourmaline quartz at the crown or root chakras to encourage integration of all aspects of physical and spiritual energies.
- Place at the thymus (upper chest) to boost the immune system and life-energy levels.

Similar stones

Rutilated quartz contains needles of gold, yellow or brown with a play of light.

Quartz with **goethite** has flat needles of a brassy colour (not black).

Actinolite quartz contains thin needles of green.

231

Clear quartz (SiO₂)

Also known as rock crystal, clear quartz was the first substance to be described as crystal. The ancient Greeks called the perfectly water-clear geometrical rock *crystalos*, meaning "permanent ice", because they thought that it was ice frozen so solidly that it wouldn't melt. By far the world's most common crystal, constituting more than a tenth of the Earth's crust, clear quartz has always been prized by healers and shamans as a means of accessing the spirit world.

Colour: transparent, white
Lustre: vitreous
Hardness: 7
System: trigonal (also recorded as hexagonal)

Identification and care

- Clear quartz crystals are quite easy to identify by their distorted, hexagonal cross-section, long prisms and six-faced termination.
- The "feel", clarity, brilliance and density varies from source to source in clear quartz.
- Tips and edges are very delicate and can easily be damaged if knocked against other stones.

A clear quartz cluster

Magic

- Amplifies the energies of other stones if placed nearby
- Helps to reveal the truth
- Used for divination and scrying

Healing functions

- Amplifies and strengthens the whole aura
- Cleanses and shifts energy
- Releases blocked emotions and helps to bring about calm
- Increases clarity of thought and sharpness of perception
- Brings spiritual peace – ideal for meditation and contemplation

Practical ideas

- To increase motivation and energy levels, and to overcome blocks to personal progress, lay out nine small clusters of clear quartz: one above the head, and the rest evenly spaced around the body, with their points facing outward. You can heighten the effect by lying on a red cloth.

Keywords
Brightening
Organizing
Amplifying

Similar stones

Danburite points are characteristically more chiselled in appearance.

Apophyllite points have four sides, not six.

Apophyllite $(KCa_4Si_8O_{20}(F.OH).8H_2O)$

This is a fairly common mineral found within cavities in volcanic rocks, in ore veins and pegmatites. It is often associated with quartz, calcite, prehnite and the zeolite family, to which it was once thought to belong. The largest source is currently Poona, India, which produces large clear crystals, as well as a beautiful green variety. The perfectly clear, brilliant pyramids that apophyllite can form make an excellent focus for cleansing and renewal processes.

Colour: clear, white, grey, green, pink

Lustre: vitreous, mother of pearl, sometimes adamantine

Hardness: 4.5–5

System: tetragonal

Identification and care
- Crystals are commonly needle-shaped, columnar or tabular, often with four sides meeting at a pyramidal point.
- Clear crystals have very shiny surfaces.
- The green variety (fluorapophyllite) is much less common.
- Handle carefully: apophyllite is a light, soft stone that is sensitive to heat.

Magic
- Develops our powers of subtle perception
- The green variety links us to nature spirits

Similar stones

Danburite is heavier and harder, with a rhombohedral section and chisel point.

Clear quartz has a hexagonal cross-section, so six faces meet at a point, not four.

Clear calcite has a less brilliant lustre.

Healing functions

- Gently cleanses our mind and body
- Helps to release feelings of insecurity and other emotions that are hard to handle
- Clusters are useful aids for meditation and encourage clarity of thought

Practical ideas

- Place apophyllite on the sacral chakra to fully harness the mineral's cleansing effect.
- Position the stone above the crown or beside the ears to encourage subtle perception and inspired thought.
- To quieten your thoughts and open your awareness to subtle levels, place the crystal on the brow.
- Gaze into apophyllite to take you beyond normal awareness of time and space.
- Place the green variety at the heart to rapidly release feelings of restriction and hopelessness.

Keywords
Nature
Clarity
Transcendence

Apophyllite from India

235

Danburite (CaB$_2$Si$_2$O$_8$)

A rare but widely distributed mineral of dolomite deposits, danburite derives its name from its main source in Danbury, Connecticut, in the USA. Other deposits are scattered around the world – notably in Mexico, Myanmar, Italy, Japan, Madagascar and Switzerland. Although occasionally golden or pink, the mineral resembles water-clear topaz. Danburite crystals are typically slender and chisel-tipped, with a sparkly brilliance and clarity. Its form suggests one of its main functions: like a chisel, it chips away at obstacles, helping us to achieve our goals.

Colour: colourless; occasionally gold or pink

Lustre: vitreous

Hardness: 7

System: orthorhombic

Identification and care
- The large, chisel-shaped danburite crystals have a frosty sparkle.
- Even though it is quite a hard stone, it is light in weight, which makes it somewhat fragile.

Magic
- Gives a positive boost to all situations
- Brings events quickly to a resolution

A single danburite crystal

Healing functions

- Activates and cleanses physical energies
- Helps us to understand the source of our emotions and how they develop
- Speeds up thought processes and boosts our intuition
- Transforms our awareness, making us more sensitive to the unseen levels of creation

Practical ideas

- For clear expression and creativity, put danburites at the base of the throat and on the sacral chakra (just below the navel).
- For protection from negativity, place selenite above the head, danburite at the throat, kunzite at the heart, carnelian at the sacral chakra and black tourmaline at the feet.
- To lift your mood, place citrine quartz at the root and solar plexus chakras, aventurine at the heart, blue lace agate at the throat and danburite at the brow.

Keywords

Brightness

Transformation

Intuition

Similar stones

Apophyllite forms pyramidal, rather than chisel-shaped, rhomboids.

Clear topaz is harder, heavier, denser and less brilliant.

Selenite (see below) is softer in radiance and texture. A mere finger-nail could scratch it.

Diamond (C)

Carbon has two very different crystal forms: the hardest known natural material, diamond, and one of the softest, graphite. Originally, diamonds were found among river and beach gravels, but now they are mined from the volcanic igneous rock kimberlite, in South Africa, Siberia and Australia. Although very hard, diamond is brittle and easily cleavable – which allows expert gemcutters to get the best from the stunning play of light that it offers. Traditionally associated with courage, purity and invincibility, diamond has always been considered powerful enough to drive away evil and bring good fortune (so long as the stone is flawless).

Colour: colourless, but also blue, green, pink, yellow

Lustre: vitreous to adamantine

Hardness: 10

System: cubic

Identification and care

- Poor-quality diamond crystals are octahedral in shape, often with convex sides. Small stones can be mounted on a crystal slice with silicon glue or beeswax to make them useable for healing.
- Cut diamonds can also be used for healing, but not if they are set in jewelry.

A cut, polished diamond

Magic
- Enables us to withstand adversity
- Signifies honour and pure intention
- Represents invincible strength

Healing functions
- Rebalances the skull, jaw and spine
- Acts as a powerful detoxifier on all levels
  – so powerful that you may need to use
  other, calming stones to modify its action
- Removes negativity and emotional blocks
- Transforms sluggish and mundane thought
  processes into a broader, universal view
- Supports meditation and spiritual ventures

Practical ideas
- Combine diamond with other stones
  to direct its brilliant light more precisely,
  for example, red stones for energizing.
- Get the most out of a small diamond
  crystal by making a gem essence.
  Simply place a few drops on the
  part of the body where you need
  diamond energy.

Keywords

Brilliance

Alignment

Connectivity

Similar stones

Cubic zirconia is heavier.

Herkimer diamond
(see below) and **spinel**
both lack the "fire" of
diamond.

Also: **clear zircon**

Ulexite (NaCaB$_5$O$_9$.8H$_2$O)

This stone forms in sedimentary borax deposits in salt lake basins and arid desert conditions. Clear crystals of ulexite are often named "TV stone" because, when placed on writing or a picture, the thin crystals act like fibre-optic cables, perfectly transferring the image to the top face of the stone as if it were a screen.

Colour: clear, white

Lustre: vitreous to silky

Hardness: 2.5

System: triclinic

Identification and care

- Ulexite can form aggregates of needle-like crystals, fibrous aggregates, granules and massive varieties.
- "TV stone" is commonly available in mineral shops. The striking transmission of light and image from lower to upper surface is unmistakeable.
- Ulexite is a soft mineral, sensitive to water and humidity.

Magic

- Facilitates immediate communication
- Helps us to weigh up opposing points of view

A slice of ulexite

240

Healing functions

- Helps the flow of neurotransmitters to all parts of the body
- Resolves blocks in our emotions
- Stimulates insight
- Provides inspiration
- Enables us to approach problems with clarity and focus
- Aids visionary experiences (soul travel, lucid dreaming, remote viewing, and so on)

Practical ideas

- Combine with blue and indigo stones to increase flow of communication or to bring about a peaceful state of mind.
- To access information about a past life, place lapis lazuli at the brow, amazonite at the throat, ulexite at the crown and smoky quartz at the feet.
- For new levels of knowledge and insight, place ulexite at the brow with fluorite on either side, kyanite at the throat, jade or watermelon tourmaline at the heart, and jet at the root chakra.

Keywords
Communication
Clarity
Visions

Similar stones

Selenite looks similar but has no "fibre-optic" effect.

Okenite has a less silky lustre and no play of light on its needles.

Zircon (ZrSiO$_4$)

This stone occurs as small pyramidal or prismatic crystals in igneous rocks, such as pegmatite. It is a hard stone, resistant to erosion, and can therefore be found in river deposits and sandstones. The colourless variety of zircon is often used as an imitation diamond or indeed, as a gemstone in its own right. Gem-quality stones tend to come from the Far East and Australia. When it comes to healing, zircon is a magical, transformative stone, with the ability to reveal hidden truths.

Colour: clear, most colours

Lustre: vitreous to adamantine

Hardness: 7.5

System: tetragonal

Identification and care

- Red-brown zircons are the most common type and are octahedral in form.
- Other coloured varieties include hyacinth, which is violet, and jargon, which is yellow.
- Zircon can be heat-treated to turn clear, blue and golden. Strongly-coloured cut zircons have most probably been heat-treated to increase their hue.
- Faceted zircon is very brilliant but care should be taken with them as the edges tend to chip easily.

A natural crystal of zircon

Magic
- Facilitates transformation
- Reveals hidden realities

Healing functions
- Cleanses and regulates the digestive and endocrine systems
- Improves our sleep
- Corrects emotional over-sensitivity
- Encourages a flexible attitude
- Reduces delusion and fantasy, and reins in the imagination
- Allows us to experience valid, practical psychic awareness

Practical ideas
- Zircon works well at the brow, crown and base of the skull.
- To increase your internal balance, place zircon on the lower abdomen.
- To improve your sense of reality, place zircon at the brow and one in each hand. If you are fearful of "losing your mind", add two more zircon – one on the top of each foot.

Keywords
Revealing
Transforming
Regulating

Similar stones

Spinel is similar, but zircon often has longer, more prismatic shapes.

Magnetite has a metallic, rather than adamantine, lustre.

243

Labradorite ((Na,Ca)Al$_{1-2}$Si$_{3-2}$O$_8$)

Labradorite is a rare feldspar named after the region of Canada where it was discovered, although it has since been found in Australia, Finland, Sicily, Norway and the Ukraine. It is noted for its stunning iridescence, caused by light bouncing off inclusions of magnetite. This unique effect is known as "labradorescence".

In healing, its iridescence acts as a reflective shield to protect the aura of the wearer. For example, any relationship problem, especially one that you can feel draining your energy reserves, will be alleviated by wearing this stone. One common side effect of the healing process is that the other person involved will experience a change in your demeanour – he or she may comment, for example, that you have become less friendly or slightly more distant. This phase will pass, however, as the energy flow between you and the other person becomes more balanced.

Colour: grey with iridescence of green, yellow, orange and peacock blue

Lustre: vitreous

Hardness: 6–6.5

System: triclinic

A piece of labradorite

Identification and care

- Pieces with little iridescence tend to be a dull grey colour and can be confused with low-grade bloodstone or beryl.

Healing functions

- Heightens our ability to focus on personal issues without letting external influences affect us
- Helps us to allow for other people's difficult emotions or moods
- Encourages us to embrace new ideas and opportunities
- Increases our ability to shift awareness outside the normal range

Practical ideas

- To release any form of negative energy from the body, including environmental stresses, place seven tumbled labradorite around the body: one at the head and one by each foot; and two evenly spaced on each side of the body. Finally, place a fluorite at the base of the throat/top of the sternum.

Keywords

Openness

Flexibility

Protection

Similar stones

Larvikite, an iridescent dark grey/black feldspar, has jigsaw-like patterning.

Spectrolite has similar iridescence, but its base colour is white or clear, whereas labradorite has a grey base colour.

245

Chalcopyrite (CuFeS$_2$)

Found all over Europe, South Africa, Australia and the Americas, chalcopyrite is a common ore of copper, which is most often displayed in its oxidized form, known as peacock ore. Unweathered crystals are a brassy colour, but once exposed to the elements, light refracts off the uneven surfaces of this soft mineral in a bewildering array of violets, purples, reds, blues, yellows and pinks. This is especially true in its massive form. Containing both copper, the metal of Venus, and iron, that of Mars, chalcopyrite represents a union of beauty and power.

Colour: brassy yellow (unweathered); blues, pinks, reds, yellows (weathered)

Lustre: metallic

Hardness: 3.5–4

System: tetragonal

Identification and care

- Chalcopyrite is usually found in its massive form; individual crystals are rare.

Magic

- Enables us to be flexible at times of change
- Fosters an appreciation of the unexpected
- Enhances creativity and innovation

A natural piece of peacock ore

Healing functions

- Stimulates the nervous system
- Removes emotional blocks, creating stable, free-flowing emotions
- Silences unwanted thoughts
- Protects us against fluctuations in energy
- Helps to ground us
- Encourages detoxification

Practical ideas

- When life seems too predictable, uninspiring and lacking in excitement, carry a piece of chalcopyrite to enliven your creative faculties and sharpen your perceptions.
- Place chalcopyrite at the brow or crown chakra to stimulate new ideas.
- Place a chalcopyrite stone at the solar plexus chakra to deepen your enjoyment of life.

Polished, metallic chalcopyrite tumble

Keywords

Protective

Cleansing

Stimulating

Similar stones

Pyrite looks similar to unweathered chalcopyrite, but it is harder and less warm in colour.

Bornite quickly tarnishes to a purplish iridescence, but is coppery red when first exposed.

Pentlandite (the main nickel ore) has granular, cubic, bronzy-brown forms. It is more earthy in colour and texture than chalcopyrite.

Opal (SiO$_{2,n}$H$_2$O)

Opal is often – unfairly – considered to be an unlucky stone. In reality, the many types of opal, including those below, offer a wide variety of healing benefits, particularly relating to the emotions. Note that opal is a delicate material, which when exposed to the air will lose water molecules, leaving small fractures.

Colour: all colours

Lustre: vitreous to mother-of-pearl

Hardness: 6

System: amorphous

Andean opal (pale blue, turquoise and green)
- Soothes the emotions
- Helps to ease relationship difficulties

Cherry opal (pink, red and orange)
- Helps to regenerate tissue and blood
- Gently increases energy levels
- Helps to lift moods

Dark opal (brown, black or grey-blue)
- Activates and balances the sacral chakra
- Eases premenstrual tension/menstrual cramps

Fire opal (rich red-orange)
- Encourages recovery after emotional upset, "burn out" or "draining" situations

Tumbled pink cherry opal

Precious opal (clear with rainbow colours)
- Helps to stabilize mood swings
- Eases the flow of life-energy through the subtle body

Common opal (no opalescence or fire; can be milky white, grey, green, purple, brown or clear)
- Provides more gentle energy than other opals
- Stabilizes the emotions
- Increases sense of self-worth

White opal (milky white with traces of other colours)
- Energizes the crown chakra
- Brings clarity of mind

Dendritic or **tree opal** (common opal with impurities that form moss-like patterns)
- Strengthens the ability to organize
- Releases constriction in the lungs, nerves and blood vessels

Precious opal in its matrix (known as boulder opal)

249

Tourmaline $(Na(Mg,Fe,Li,Mn,Al)_3Al_6(BO_3)Si_6.O_{18}(OH,F)_4)$

Almost every colour and every combination of colour can be found within the tourmaline group of crystals. Unfortunately, tourmaline tends to crystallize in hard rocks like granite or quartz (see pages 230–231), which makes it difficult to extract the long, thin crystals without breaking them. Thus, tumbled pieces and short sections are often inexpensive, whereas complete terminated wands are much more difficult to find, and if they display a range of colours within the transparent crystal they can command very high prices.

Below are some notable colour varieties of tourmaline. Regardless of their colour, all tourmalines with striations along the length of their crystals speed energy flow, helping to clear blockages and restore equilibrium.

Colour: all colours

Lustre: vitreous

Hardness: 7–7.5

System: trigonal

Indicolite (blue)

- Balances the throat chakra, so benefits the thyroid, lungs, larynx and bones of the neck
- Activates the brow chakra, maximizing perception, intelligence and absorption of information

Rubellite (red or pink)

- Can be a stimulating or calming stone, depending on the depth of colour
- Helps to correct excessively aggressive or passive personalities
- Energizes the sacral chakra, which increases all forms of creativity

Watermelon tourmaline
(green intermixed with pink)
- Excellent for balancing the heart chakra on all levels

Verdelite (green)

- Boosts the immune system
- Realigns bones and strained muscle tissue
- Increases confidence and helps to give us a sense of belonging

Other tourmalines include: **achroite** (clear); **elbaite** (multicoloured); **dravite** (brown); **siberite** (violet); **uvite** (black, brown or yellow); and **schorl** (black, see pages 48–49).

A slice of watermelon tourmaline from Brazil

Coral (CaCO₃)

Formed from the skeletal remains of colonies of small sea animals and polyps, coral is usually found in shallow, warm water. Red varieties are the most valuable and are collected off Japan, Africa and Malaysia, and in the Mediterranean and Red Sea. Australia and the Caribbean yield black and golden coral; while blue coral is found off Cameroon in West Africa. In Tibet and China fossilized red coral can be found at the top of mountains, along with fossilized conch shells, testifying to the long history and ever-changing surface of the planet. Although not strictly a mineral unless it has become fossilized, coral has been a popular gemstone for thousands of years. The red variety, in particular, is cherished in India, Tibet and among Native Americans as a promoter of fertility and life-energy.

Colour: red, white, pink, gold, blue, black

Lustre: dull to vitreous

Hardness: 3

System: hexagonal or trigonal

Identification and care

- Fossilized coral is heavier than unfossilized coral.
- Antique beads and fossilized coral are more ethical purchases than live coral, which is endangered.

Tumbled blue coral

- Although not particularly soft, thin strands of coral are brittle and should be handled and stored with care.

Magic
- Heightens awareness of the needs of others
- Helps us to understand how all life connects
- Protects us from harm

Healing functions
- Strengthens the heart, circulation and bones
- Balances and steadies the emotions
- Boosts fertility and our drive to satisfy desires

Practical ideas
- To gently stimulate and warm the circulatory system, place red or pink coral at the root, sacral and brow chakras.
- To increase fertility, place red coral at the root chakra, pink at the sacral chakra, opal at each hip and smoky quartz between the feet.

Keywords

Community

Strength

Connectivity

Similar stones

Imitated by plastic and glass, coral can be identified by its "grain" of streaks and dots.

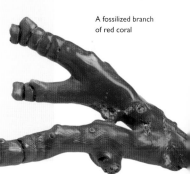

A fossilized branch of red coral

Pearl (CaCO$_3$ + conchiolin + H$_2$O)

Perfect balls of shimmering iridescence, good-quality pearls are hard to find. Like all things of great value, they therefore have to be diligently sought out. Pearls form inside the shells of fresh- or sea-water pearl oysters where there is irritation from grains of sand or parasites. Just as beautiful is mother-of-pearl, which is the inside of the actual shell of the pearl oyster. Abalone and paua are similar shells. Like all gems that come from the sea, pearl relates in particular to the flow of emotions, intuition and the power of the moon.

Colour: white, pink, brown, black, blue, green

Lustre: mother-of-pearl

Hardness: 2–2.5

System: amorphous, orthorhombic

Identification and care

- Irregularly shaped freshwater pearls are more affordable than perfectly formed ones.
- Pearl and related shells are soft, so should be stored separately.

Magic

- Calms extreme situations
- Encourages flow and change

254

Natural sea-water pearls

Healing functions

- Regulates glandular function
- Optimizes assimilation of nutrients
- Balances the emotions, increasing tolerance and amenability
- Energizes the sacral chakra
- Reduces worry, anxiety and frustration
- Helps us to focus on the core of the Self: to understand who we are and who we are not

Practical ideas

- To loosen stiff hip joints or to ease lower back pain: place pearl at the sacral chakra; lodestone at the base of the neck; and black tourmaline above the head and below the feet.
- To release stress and worry (or to overcome addictive tendencies): place pearl at the solar plexus; clear quartz above the head; and smoky quartz or any other grounding stone between the feet.

Keywords
Sensitivity
Tolerance
Alignment

Similar stones

Moonstone is much harder and has less colour than pearl.

Abalone and **paua** have greater rainbow iridescence and shine than pearl. Abalone is less silvery and paua has more blues, greens and pinks in its iridescence.

255

Pietersite $(SiO_2+Na_2(Mg,Fe,Al)_5(OH/Si_4O_{11})_2)$

A recently discovered mineral, pietersite is
an unusual mix of the earthy and the celestial.
The stone often looks like a storm in a teacup,
a swirl of contrasting colours in turbulent
conflict – hot and cold, fire and rain all mixed
up together with a cool, almost metallic, lustre.
Pietersite derives from layers of sand or silt
that have become cemented together by
quartz in the form of blue falcon's eye and
yellow-gold tiger's eye (see pages 70–71).
Rather than having clearly defined parallel
fibres, in pietersite these components are
compressed and mixed in random ways to
create a uniquely energetic stone.

Colour: blue-black with brown-gold flecks	
Lustre: vitreous	
Hardness: 7	
System: trigonal	

Identification and care

• The combination of golden browns, cool
  blues and white – all in a dynamic sinuous
  pattern – easily identifies this stone.

Magic

• Provides creative ideas to bring drifting,
  stagnant situations to a resolution

• Helps us to resolve conflict

Similar stones

Bronzite lacks the blue
veils to be found in
pietersite.

Healing functions

- Relaxes tense muscles in the middle back
- Helps to release negative feelings about a past event or situation without us having to go through the pain of reliving it
- Promotes a sense of detachment, so that we can process and integrate thoughts in our own way
- Provides an anchor during times of spiritual transformation

Practical ideas

- If you feel that you are stuck in a difficult situation, place some pietersite at the sacral chakra, just below the navel, and put a tiger's eye at the solar plexus and an amethyst at the brow.
- Place a piece of pietersite at the brow to initiate new ways of seeing old situations.
- Place a pietersite stone at the solar plexus to review how you see yourself.

Keywords

Stirring

Clearing

Anchoring

Tumbled pietersite

257

Unakite $((Ca_2(Al, Fe^{3+})_3SiO_4)_3(OH)/Ca.Al_2Si_2O_8)$

Composed of a mixture of pink feldspars, green epidote and quartz, unakite is very similar in structure to granite. The result is an opaque, pastel mix with contrasting areas of pink and green. This combination of colours works particularly well with the energies of the heart, where the pink strengthens our sense of identity and helps to release emotional stresses and the green encourages a sense of expansion and growth in a calm and balanced way.

As a rock of massive, opaque form, unakite prevents us from getting too wrapped up in our emotions, lending a practical, down-to-earth energy to any healing process in which it is used.

Colour: pink and green	
Lustre: vitreous	
Hardness: 6–7	
System: none	

Identification and care

- Unakite is an opaque rock containing areas of pink feldspar and pale green epidote.
- It is usually seen in the form of beads, cabochons or tumbled stones.

A unakite palmstone

Magic

- Helps to resolve conflicts
- Loosens constrictions that we might feel in our lives

Healing functions

- Balances physical aspects of the heart and circulation
- Soothes any aches and pains in the pelvic area
- Enhances feelings of self-worth
- Gently releases emotional blocks
- Focuses attention on the present, helping us to distance ourselves from past difficulties
- Encourages us to take a detached view of our lives, problems and accomplishments

Practical ideas

- To calm flyaway emotions, resolve interpersonal issues, or to help you develop a more flexible attitude: place one unakite stone at the solar plexus and one at the heart centre, adding a grounding stone below the feet.

Keywords

Perspective

Balance

Self-worth

Similar stones

Pink **granite** lacks green areas, and its crystals refract more light.

Ruby in **zoisite** (see below) has a deeper green/red colour mix and more clearly defined areas of colour.

259

Fossils

The petrified remains of the most ancient life
forms on this planet, fossils amalgamate the
energies of animal, vegetable and mineral.
Fossils are formed when dead plants or animals
become covered by layers of sediment, almost
always in water. Over time, in this oxygen-free
environment, the organic material is replaced
by layers of mineral or dissolves completely
to leave a hollow cast, which often fills with
another mineral.

Many fossils with striking patterns and
shapes have been regarded as protective and
powerful talismans for thousands of years. All
fossils, regardless of their appearance, can be
used in healing to:

● Stabilize, ground and steady the life-force
● Restore order and pattern to
  disharmonious states
● Access distant memory and travel through
  time and space
● Understand the vast continuity of life

Individual types of fossil, such as those
listed on the opposite page, also have their
own specific healing capabilities.

Ammonite

Ammonite (the shell of an ancient sea
creature forming a slowly expanding spiral
with complex surface patterns)
- Strengthens the chakra system
- Helps us to accept change and cycles of time

Sea urchin (a heart-shaped fossil, also known
as a "sand-dollar", with a five-pointed star on top)
- Encourages coordinated activity
- Protects delicate sensibilities

Dinosaur bone (fragments of these once
dominant giant reptiles)
- Brings assuredness and confidence
- Gives security and strength
- Enables us to control circumstances

Fossil plant (imprinted plant matter)
- Provides us with the continued
  sustenance of the sun's energy
- Stabilizes the solar plexus
- Brings contentment
- Enables us to absorb life-energy
- Gives us "breathing space"

Fossil tree bark

Rocks and pebbles

Unlike crystals, which tend to have a uniform chemical structure (although sometimes with inclusions of other minerals), rocks are an amalgamation of many minerals. All rocks have a stabilizing and normalizing effect. Below are common rocks that can be used in healing.

Granite

Granite (a granular and shiny igneous rock, composed of quartz, feldspar and mica)
- Balances subtle bodies and speeds healing processes (especially red and pink varieties). Place pieces by the throat and feet and hold in each hand.

Limestone (a pale, compact rock of calcium carbonate, formed from calcite-rich shells and corals deposited in ancient warm seas, it forms landscapes with cliffs, gorges and cave systems)
- Energizes the root and sacral chakras, stimulating creativity and sensitivity
- Stabilizes the emotions and reduces tension and fear
- Works well when placed at the sacral chakra or the base of the skull

Limestone

Sandstone

Sandstone (formed when sand is cemented
with silica or calcite)
- Helps to maintain elasticity and flexibility
  of tissues like blood vessels and skin

Basalt (a grey-green to black, fine-
grained rock with bubbles or crystal
cavities, which is ejected as lava from
volcanoes)
- A solid, grounding stone for steadiness and
  maintenance of energy levels

Basalt

Chalk (a white, soft, porous rock often found
with flint and fossils, it is formed from
deposits of microscopic marine organisms)
- Protects the aura from negativity and
  disruptive influences

Chalk

Marble (metamorphosed limestone, veined
and banded in many colours)
- Very cooling and calming
- Stabilizes the energies of the solar
  plexus chakra
- Balances extremes of behaviour

Marble

263

Natural forms of quartz

As one of the most abundant minerals on the planet, quartz can be found in a wide range of shapes, depending on the manner of its formation. Each of these forms – notable examples of which are described below – brings distinct healing qualities.

Double termination (a crystal with a natural termination at each end – occurs when the crystal has grown in soft or liquid conditions)
- Integrates and unifies
- Promotes energy flow within the body
- Can be held in both hands to aid meditation

Double termination

Herkimer diamond (a type of small, brilliant double-terminated quartz first found in soft rock in Herkimer County, New York, USA)
- Detoxifies at all levels
- Encourages lucid dreaming, especially if placed under the pillow

Elestial quartz (often smoky quartz with mica inclusions, it has multiple terminations, giving it a stepped or scaly appearance)

Herkimer diamond

- Excellent for deep healing
- Leads us into profound levels of awareness
- Grounds, clears and integrates energy

Phantom quartz (contains a fine shadow, or phantom – the vestige of an earlier phase in the crystal growth)
- Helps us to understand hidden causes
- Provides a good focus for meditation

Phantom quartz

Tabular quartz (a thin, flat form of quartz)
- Allows a frictionless flow of energy

Laser wand (tapering) (a natural quartz point where the often slightly curved sides narrow considerably toward the termination)
- Focuses and concentrates energy – excellent for precise or deep body healing
- Expands awareness

Laser wand (widening) (a rarer form where the wand widens toward the tip)
- Releases and expands energy, promoting a smooth flow – much like tabular quartz

Laser wand

265

Self-healed quartz (a crystal that has broken apart during formation and then grown back together – the joint is identifiable by a cloudy area and a distinct change of axis)
● Helps us to understand the cause of a disease
● Accelerates repair processes

Generator quartz (has six identical faces meeting at the apex)
● Energizes its surroundings
● Provides a focus for healing
● Ideal for group work or as a focus for personal meditation

Self-healed quartz

Channelling quartz (the largest face of this crystal has seven sides opposite which is a smaller, triangular face)
● Ideal for use in meditation
● With the large face held at the brow, it can help intuition and channelling of energy

Transmitter quartz (has two seven-sided faces with triangular faces between them in a geometric combination of three and seven)

Channelling quartz

- Receives and transmits information from the aura: after grounding personal energy, a clearly defined question can be projected into the crystal. Leave it in a quiet place for at least 12 hours and then return to it, placing one triangular facet to your brow to absorb relevant information.

Record keeper quartz (one or more faces are smooth except for a raised or indented distinct pyramid shape on the surface)
- Harmonizes and integrates
- The etched shape activates fine levels of the mind and memory
- Ideal for personal meditation, it can be held in the hands and then placed at the brow in order to unlock aspects of self-knowledge

Cathedral or **Candle quartz** (looks like several quartz crystals melded together, buttressing each other – it usually has one main point, with other terminations surrounding it)

Transmitter quartz

- Cathedral quartz is thought to give access to a body of shared knowledge, sometimes known as the "Akashic Records", held within the collective mind of humanity
- Ideal as a personal or group meditation focus
- Helps us to determine a personal spiritual path or direction

Window quartz (a crystal with prominent diamond-shaped faces situated between the larger faces)

- Helps with health assessments – a healer scans over a person's aura with the crystal then holds it to his or her (the healer's) brow to receive relevant information

Window quartz

Time-link quartz (a type of window quartz that has slanted diamond or parallelogram windows)

- Shifts awareness to different times, places and dimensions
- If the crystal slants to the left, it is thought to reveal past lives; if to the right, our potential future

Time-link quartz

Isis quartz (the largest face is five-sided)
- Named after the Egyptian goddess Isis, who laboriously worked to find the scattered remains of her dead husband, this crystal encourages us to strive for success and create order out of chaos
- Heals damaged emotions
- Enables us to help others without becoming too involved with their problems

Lemurian seed quartz (dull in appearance and slightly pink, with noticeable horizontal striations running along one or two sides of the crystal)
- Has a different kind of energy from other quartz, which can make it disconcerting to hold at first
- Energizes connections between the Earth and the rest of the universe
- Helps us to perceive our position within the greater scheme of things

Lemurian seed quartz

Specially cut quartz

Although crystals form naturally in all manner of
shapes and sizes, some crystal healers have cut
their own shapes for specific healing purposes.

Magician crystal

This is simply a double-pyramid cut stone
whose shape links to all the elements. A
traditional ceremonial magician would work
with the four elements while contacting both
the universe and the Earth.

Vogel crystals

These are named after Marcel Vogel, a scientist
and metaphysician who investigated how crystal
interacted with the subtle layers of the body
and who designed crystal tools to work on
different levels of energy.

His healing crystal is a double-terminated
wand, one end wider than the other. The facets
encourage the flow of energy through the
crystal. The crystal's function can be enhanced
through tightly focused thought or intention.
Vogel also devised a meditation crystal, which
is single-terminated and shaped like an obelisk.

A healing Vogel crystal

Merkaba crystal

The merkaba is cut into two interpenetrating pyramids (double tetrahedron). It symbolizes the link between heart, mind and body, encouraging the easy flow of energy and information between all levels of being.

Platonic solids

Quartz is sometimes cut into five shapes, the "Platonic solids", thought to represent the primary patterns of matter. They are also linked to elements and chakras (see table below).

Shape	Faces	Element	Chakra
Tetrahedron	4	Fire	Solar plexus
Cube	6	Earth	Root
Octahedron	8	Air	Heart
Dodecahedron	12	Ether	Throat
Icosahedron	20	Water	Sacral

To balance the chakras, place the five Platonic solids on the appropriate chakras, along with a merkaba crystal on the brow chakra and a spherical crystal on the crown chakra.

Artificial quartz

Nowadays virtually all natural crystals have their artificial counterparts. Even for jewelers it is becoming very difficult to distinguish one from the other.

Artificial quartz has a particularly long history, having been first created synthetically in 1845. However, it was not until the 1970s that technology allowed the creation of large enough amounts to make it commercially viable. Solutions of sodium carbonate or sodium hydroxide are subjected to high temperature and pressure. The solution is carried across seed crystals, which then grow.

Although natural quartz has stronger energy, and so is normally better to use in healing, synthetic quartz still has an internal lattice structure, which some believe can bring healing benefits. In contrast, glass, although it has the same chemical formula as quartz (SiO_2), has no internal crystal structure at all. This means that artificial crystals made of glass, such as Austrian crystal, are of little use in healing.

Laboratory-created quartz crystals, such as the examples listed opposite, are often vivid and eye-catching. The colour of the stone dictates the chakra with which it will correspond.

Siberian gold quartz (solar plexus chakra)
- Balances polarities
- Confers inner wisdom

Siberian green quartz (heart chakra)
- Calms and balances

Siberian blue quartz (brow chakra)
- Builds intuition, perception and psychic skills

Siberian purple quartz (crown chakra)
- Transforms energy, releases energy blocks

Pink lazurine
(sacral chakra)
- Changes negative to positive
- Helps us to address "self" issues, such as low levels of self-love and self-tolerance

Pink lazurine (front) and blue obsidian (back), another commonly man-made stone 273

Enhanced quartz

As well as completely synthetic quartz, there
is also enhanced quartz – natural quartz that
has been treated in some way to alter its
appearance and healing properties.

One of the most common procedures
is metal-coating, in which clear quartz points
are placed in a pressurized vacuum chamber
at very high temperature and exposed to
vaporized metal. This creates a permanent layer
a few atoms thick on the surface. Quartz can
also be dyed striking colours. The following are
common examples of enhanced quartz.

Aqua aura (clear quartz infused with gold
vapour to create a rich, iridescent blue surface)
- Harnesses the power of gold and quartz
  working together
- Stimulates the finer levels of thought and
  communication

Rainbow aura (clear quartz treated with
vaporized platinum, creating a rainbow-like
opalescence; also known as opal aura, pearl
aura and angel aura)

274

Rainbow aura

- Helps to clear the mind for meditation and for fine levels of communication
- Linked to the crown chakra, so will energize all subtle systems

Ruby aura (clear quartz bonded with vaporized gold and platinum to create a rich red hue)
- Grounds and protects
- Linked to the root chakra, so helps the back, spine, hips, legs and feet

Sunshine aura (clear quartz treated with platinum and gold, giving a bright yellow colour)
- Linked to the solar plexus chakra, enhancing digestion and emotional stability

Crackled (or star-burst) quartz
(clear quartz stressed by heating and cooling then dyed in a range of colours that are absorbed into the fractures thus formed, giving a very sparkly appearance)
- Affects the chakra that corresponds to its colour (see page 15)

Crackled quartz

275

For reference

Glossary

adamantine diamond-like, very hard

agate mixture of cryptocrystalline or microcrystalline chalcedony, rhombohedral quartz and shapeless opal in close bands

aggregate collection of more than one mineral

allochromatic coloured by impurities

asterism star-light effect appearing on a crystal surface caused by internal microcrystals

Atlantis mythical continent in the mid-Atlantic Ocean

aura general term for the electro-magnetic field around a body or object

axis imaginary straight line of rotation

botryoidal resembling a bunch of grapes

cabochon method of cutting gemstones to give a flat base and domed surface

chakra spinning vortex of subtle energy

chalcedony any of a group of microcrystalline varieties of quartz

chatoyancy changeable, undulating lustre, like a cat's eye

ch'i Chinese term for life-energy

cleavage plane along which a crystal can be split

columnar in the form of a column

compound combination of elements

concretion mass of mineral found in rock

cryptocrystalline crystalline structure that is concealed

crystal mineral with regular planes and faces reflecting a regular internal structure

crystal system characteristic grouping of atoms defined by angles of axes and symmetry

dendritic having tree-like markings

dichroism showing one of two colours depending on angle of view

element single chemical or substance of which matter is formed

feldspar any of a group of aluminosilicates often occurring as large, pale crystals

ferruginous of, or containing, iron

gneiss metamorphic rock containing quartz and feldspar

grounding stone any crystal that focuses us in the here and now

hydrothermal relating to heated water

igneous relating to rock produced by the action of fire or volcano

inclusion substance contained in a crystal or mineral

iridescence interchange of colours

lattice regular pattern of atoms within a crystal structure

macrocosmic relating to the universe

massive relating to a mineral that is not definitely crystalline

matrix base rock on which crystals grow

metamorphic relating to rock modified by heat and pressure

microcosmic relating to the world of human nature

microcrystalline having crystals too small to be seen by the naked eye

mineral single chemical compound

Mohs scale mineral hardness scale devised by Friedrich Mohs

nadi energy channel in the subtle body

occiput back of the head at the top of the neck

pegmatite any of a group of crystallized granites containing mica

periodic table list of all chemical elements in order of atomic weight

phantom ghost-like form within a crystal

pleochroism showing one of more than two colours depending on angle of view

polarization modification of light depending on angle of view

prana Indian term for life-energy

pyroxene any of a group of minerals composed of silicates of iron, calcium and magnesium and found within igneous rock

refraction diversion of light from its natural course

rock combination of minerals and crystals

schiller twinkle or sparkle created by light reflecting off microcrystals

schist crystalline rock where the minerals are arranged in parallel layers

scrying perceiving what is hidden, usually using a crystal ball or mirror

sedimentary relating to rock formed from layers compressed by its own weight and water

striated having a grooved surface

subadamantine dulled diamond-like sparkle

subtle bodies seemingly discrete layers of the body's electromagnetic field

subvitreous not quite vitreous

tabular slab-like or with a flat base and top

termination pointed tip of a crystal

vitreous resembling glass

Mineralogical information

The Mohs scale of relative hardness

The "hardness" scores in the Guide to Crystals come from a scale devised in 1812 by the German mineralogist Friedrich Mohs to assist in identifying different minerals. To use the scale (see below), you will need an example of each of the ten "test" stones. You can work out the hardness of an unknown mineral, and so get a better idea of its identity, by scratching an inconspicuous or damaged part of the stone with the test stones. For example, if a crystal is capable of scratching fluorite, but is scratched by apatite, then you can conclude that it has a hardness of between 4 and 5 on the Mohs scale.

1 (softest) .... talc	6 .... orthoclase
2 .... gypsum	7 .... quartz
3 .... calcite	8 .... topaz
4 .... fluorite	9 .... corundum
5 .... apatite	10 (hardest) .... diamond

Chemical symbols used in "A Guide to Crystals"

Ag silver	**Ca** calcium	**K** potassium	**S** sulphur
Al aluminium	**Cl** chlorine	**Li** lithium	**Si** silicon
Au gold	**Cr** chromium	**Mg** magnesium	**Sr** strontium
B boron	**Cu** copper	**Mn** manganese	**Ti** titanium
Ba barium	**F** fluorine	**Na** sodium	**Zn** zinc
Be beryllium	**Fe** iron	**O** oxygen	**Zr** zirconium
C carbon	**H** hydrogen	**P** phosphorus	

Bibliography

Geinger, Michael, *Crystal Power, Crystal Healing*, Cassell & Co. (London), 1998

Gerber, Richard, *Vibrational Medicine*, Bear & Co. (Santa Fe, New Mexico), 1988

Gurudas, *Gem Elixirs and Vibrational Healing 1*, Cassandra Press (Boulder, Colorado), 1989

Hankin, Rosie (ed.), *Rocks, Crystals, Minerals*, New Burlington (London), 1998

Korbel, Petr and Milan Novak, *The Complete Encyclopedia of Minerals*, Rebo (Lisse, Netherlands), 1999

Kourimsky, Dr J., *The Illustrated Encyclopedia of Minerals and Rocks*, Aventinum (Prague), 1977

Lilly, Simon, *The Illustrated Elements of Crystal Healing*, Harper Collins/Element (London), 2000 and 2002

Lilly, Simon and Sue, *Crystal Doorways*, Capall Bann (Milverton, Somerset), 1997

Lilly, Sue and Simon, *Colour Healing*, Lorenz (London), 2001

Lilly, Sue and Simon, *Crystal Healing*, Lorenz (London), 2001

Lilly, Sue, *Crystal Decoder*, Barron's (USA) and Fair Winds (UK), 2001

Raphael, Katrina, *Crystal Enlightenment*, Aurora (New York), 1985

Crystal healing associations

UK associations

Affiliation of Crystal Healing Organizations
www.crystal-healing.org

Crystal and Healing Federation
www.crystalandhealing.com

Institute of Crystal and Gem Therapists
www.greenmantrees.demon.co.uk

US associations

Association of Melody Crystal Healing Instructors
www.taomchi.com

Crystal Academy of Advanced Healing Arts
www.crystalacademy.cncfamily.com

Index

Acknowledgments

Authors' acknowledgments

We would particularly like to thank Alex Mason at Crystals (3 Flexi Units, Budlake Road, Exeter, EX2 8PY; www.crystalshop.co.uk) for sourcing and lending such beautiful crystals to be photographed in this book.

We would also like to thank Angela Andrew, Terrie Lockyer, Carol Saunders, Khalid Haq, Veronica Ditchfield, Helen Jones, Luciana Corp, Ken and Mandy Robinson and Rosanna Stancampiano for their help and input in discerning the properties of Preseli bluestone.

Resources
Crystals (www.crystalshop.co.uk)
Preseli Bluestone Limited (www.stonehengestones.com)
British Association of Flower Essence Producers (www.bafep.com)
Green Man Essences (www.greenmanessences.com)

Publisher's acknowledgments

The publisher would like to thank the following people and photographic libraries for permission to reproduce their material:
Page 10 Jürgen Liepe, Berlin/Egyptian Museum, Cairo; **19** Photolibrary.com/ Dynamic Graphics (UK) Ltd; **31** Simon and Sue Lilly; **238** Getty Images/ Photographer's Choice/Steve Cole